China-ASEAN Clean Energy Capacity Building Programme

Research on Applications of Low Wind Speed Power in ASEAN

China Renewable Energy Engineering Institute

· Beijing ·

Abstract

As one of the China-ASEAN Clean Energy Capacity Building Programme Technical Materials Publication Series, the book provides insights into the status quo and trend in global wind power and low wind speed technology (LWST) development and analyzes the key LWST challenges. It focuses on ASEAN, assesses the onshore wind potential of the ASEAN member states, and offers suggestion on the major layout, priorities, and potential challenges and solutions for wind power development in the region. This book will serve as a valuable reference in the efforts to facilitate LWST cooperation and development and push forward clean energy transition of the region.

The book is customized for experts specializing in wind power technology R&D and resource assessment and planning, practitioners of international engineering businesses, and the faculty and students of the related disciplines in colleges and universities.

图书在版编目（CIP）数据

中国-东盟清洁能源能力建设计划 ：东盟国家低风速风电应用潜力研究 = China-ASEAN Clean Energy Capacity Building Programme: Research on Applications of Low Wind Speed Power in ASEAN ：英文 / 水电水利规划设计总院编著. -- 北京 ：中国水利水电出版社，2022.12
ISBN 978-7-5226-1158-7

Ⅰ. ①中… Ⅱ. ①水… Ⅲ. ①风力发电－无污染能源－能源发展－研究－中国、东南亚国家联盟－英文 Ⅳ. ①F426.2②F433.62③TM614

中国版本图书馆CIP数据核字(2022)第242821号

书　　名	China - ASEAN Clean Energy Capacity Building Programme **Research on Applications of Low Wind Speed Power in ASEAN**
中文书名拼音	DONGMENG GUOJIA DIFENGSU FENGDIAN YINGYONG QIANLI YANJIU
作　　者	China Renewable Energy Engineering Institute
出版发行	中国水利水电出版社 （北京市海淀区玉渊潭南路1号D座　100038） 网址：www. waterpub. com. cn E - mail：sales@mwr. gov. cn 电话：(010) 68545888（营销中心）
经　　售	北京科水图书销售有限公司 电话：(010) 68545874、63202643 全国各地新华书店和相关出版物销售网点
排　　版	中国水利水电出版社微机排版中心
印　　刷	天津嘉恒印务有限公司
规　　格	184mm×260mm　16开本　7.5印张　245千字
版　　次	2022年12月第1版　2022年12月第1次印刷
印　　数	001—700册
定　　价	**100.00**元

China-ASEAN Clean Energy Capacity Building Programme
Technical Materials

Project Steering Committee

“中国-东盟清洁能源能力建设计划系列技术材料

专项领导小组

组　长

李　昇	水电水利规划设计总院院长
易跃春	水电水利规划设计总院常务副院长
顾洪宾	水电水利规划设计总院副院长
张益国	水电水利规划设计总院总规划师
Dr. Nuki Agya UTAMA	东盟能源中心执行主任

成　员

姜　昊	水电水利规划设计总院 国际业务部主任
张木梓	水电水利规划设计总院 国际业务部综合处处长
魏颖婕	水电水利规划设计总院 国际业务部综合处分析员、翻译
Beni SURYADI	东盟能源中心 可持续能源和可再生能源部代理经理、电力、化石燃料、替代能源和储能部经理

The Compilation Organizations and Group of *Research on Applications of Low Wind Speed Power in ASEAN*

COMPILATION ORGANIZATIONS

China Renewable Energy Engineering Institute

ASEAN Centre for Energy

POWERCHINA Chengdu Engineering Corporation Limited

POWERCHINA Zhongnan Engineering Corporation Limited

POWERCHINA Kunming Engineering Corporation Limited

POWERCHINA Hubei Electric Engineering Co., Ltd.

Xinjiang Goldwind Science and Technology Co., Ltd.

Zhejiang Windey Co., Ltd.

Envision Group

Chinese Renewable Energy Industries Association

Global Energy Interconnection Development and Cooperation Organization

COMPILATION GROUP

Edited by

Dr. LI Sheng	Director General China Renewable Energy Engineering Institute
YI Yuechun	Executive Director General China Renewable Energy Engineering Institute
Dr. GU Hongbin	Deputy Director General China Renewable Energy Engineering Institute
ZHANG Yiguo	Chief Planning Engineer China Renewable Energy Engineering Institute
Dr. Nuki Agya UTAMA	Executive Director ASEAN Centre for Energy

Reviewed

XIE Hongwen	Deputy Chief Engineer China Renewable Energy Engineering Institute
Dr. JIANG Hao	Director of International Cooperation Department China Renewable Energy Engineering Institute
ZHANG Shishu	Deputy Secretary of the Party Committee and Chairman of the Trade Union POWERCHINA Chengdu Engineering Corporation Limited
YAO Xiyu	Vice President of New Energy Engineering Design Institute POWERCHINA Zhongnan Engineering Corporation Limited
TANG Yan	Advisor POWERCHINA Hubei Electric Engineering Co., Ltd.
LU Min	Deputy Director of Pumped Storage Survey and Design Management Department POWERCHINA Kunming Engineering Corporation Limited
MIAO Mulu	Deputy Director of New Energy Engineering Design Institute Office POWERCHINA Zhongnan Engineering Corporation Limited
JIANG Jianhong	Chief Engineer of New Energy Engineering Company POWERCHINA Chengdu Engineering Corporation Limited
CHENG Jie	Deputy General Manager and Chief Engineer of Planning Company POWERCHINA Hubei Electric Engineering Co., Ltd.

Compiled by

(China)

ZHANG Muzi	China Renewable Energy Engineering Institute
XIA Ting	China Renewable Energy Engineering Institute
DENG Zhenchen	China Renewable Energy Engineering Institute
WEI Yingjie	China Renewable Energy Engineering Institute
WANG Yicheng	China Renewable Energy Engineering Institute
LIU Zhenyang	China Renewable Energy Engineering Institute
CHEN Wen	POWERCHINA Zhongnan Engineering Corporation Limited
LI Dan	Chinese Renewable Energy Industries Association
CHEN Xu	POWERCHINA Hubei Electric Engineering Co., Ltd.

QIU Xin	POWERCHINA Chengdu Engineering Corporation Limited
YU Yongjing	POWERCHINA Chengdu Engineering Corporation Limited
YU Dingkun	POWERCHINA Chengdu Engineering Corporation Limited
QIN Junjie	POWERCHINA Zhongnan Engineering Corporation Limited
YANG Xueqian	China Renewable Energy Engineering Institute
LI Le	China Renewable Energy Engineering Institute
CHEN Juanjuan	China Renewable Energy Engineering Institute
XIE Deng	POWERCHINA Chengdu Engineering Corporation Limited
ZHU Mingliang	Xinjiang Goldwind Science and Technology Co. , Ltd.
CHEN Yang	Xinjiang Goldwind Science and Technology Co. , Ltd.
SHEN Xinhe	Zhejiang Windey Co. , Ltd.
WANG Haijiang	Xinjiang Goldwind Science and Technology Co. , Ltd.
WANG Yunyan	Envision Group
WU Jiawei	Global Energy Interconnection Development and Cooperation Organization

(ASEAN)

Beni SURYADI	ASEAN Centre for Energy
Akbar Dwi WAHYONO	ASEAN Centre for Energy
Gabriella IENANTO	ASEAN Centre for Energy
Adhityo Gilang BHASKORO	ASEAN Centre for Energy
Rifa FADILLA	ASEAN Centre for Energy
Annisa Sekar LARASATI	ASEAN Centre for Energy
Raisha VERNIASTIKA	ASEAN Centre for Energy
Dr. Akbar SWANDARU	ASEAN Centre for Energy
Dr. Zulfikar YURNAIDI	ASEAN Centre for Energy
Monika MERDEKAWATI	ASEAN Centre for Energy
Dynta Trishana MUNARDY	ASEAN Centre for Energy

Translation Reviewed by

WEI Yingjie	China Renewable Energy Engineering Institute

Translated by

WANG Yicheng	China Renewable Energy Engineering Institute
YANG Xueqian	China Renewable Energy Engineering Institute

《东盟国家低风速风电应用潜力研究》编写单位及成员名单

编　写　单　位

水电水利规划设计总院
东盟能源中心
中国电建集团成都勘测设计研究院有限公司
中国电建集团中南勘测设计研究院有限公司
中国电建集团昆明勘测设计研究院有限公司
湖北省电力勘测设计院有限公司
新疆金风科技股份有限公司
浙江运达风电股份有限公司
远景能源有限公司
中国循环经济协会可再生能源专业委员会
全球能源互联网发展合作组织

编写组成员名单

主　编

李　昇　水电水利规划设计总院院长
易跃春　水电水利规划设计总院常务副院长
顾洪宾　水电水利规划设计总院副院长
张益国　水电水利规划设计总院总规划师
Nuki Agya UTAMA　东盟能源中心执行主任

校　审

谢宏文　水电水利规划设计总院副总工程师
姜　昊　水电水利规划设计总院国际业务部主任
张世殊　中国电建集团成都勘测设计研究院有限公司党委副书记、工会主席、职工董事

姚曦宇　中国电建集团中南勘测设计研究院有限公司新能源工程设计院副院长
唐　焱　湖北省电力勘测设计院有限公司调研员
卢　敏　中国电建集团昆明勘测设计研究院有限公司抽水蓄能勘测设计管理部副主任
苗沐露　中国电建集团中南勘测设计研究院有限公司新能源工程设计院室副主任
蒋建红　中国电建集团成都勘测设计研究院有限公司新能源工程分公司总工程师
程　杰　湖北省电力勘测设计院有限公司规划分院副总经理兼总工程师

编写成员

（中国）

张木梓　水电水利规划设计总院
夏　婷　水电水利规划设计总院
邓振辰　水电水利规划设计总院
魏颖婕　水电水利规划设计总院
王艺澄　水电水利规划设计总院
刘镇洋　水电水利规划设计总院
陈　文　中国电建集团中南勘测设计研究院有限公司
李　丹　中国循环经济协会可再生能源专业委员会
陈　旭　湖北省电力勘测设计院有限公司
仉　欣　中国电建集团成都勘测设计研究院有限公司
郁永静　中国电建集团成都勘测设计研究院有限公司
俞定坤　中国电建集团成都勘测设计研究院有限公司
覃俊杰　中国电建集团中南勘测设计研究院有限公司
杨雪倩　水电水利规划设计总院
李　乐　水电水利规划设计总院
陈娟娟　水电水利规划设计总院
谢　登　中国电建集团成都勘测设计研究院有限公司
朱明亮　新疆金风科技股份有限公司
陈　杨　新疆金风科技股份有限公司
申新贺　浙江运达风电股份有限公司

王海江　新疆金风科技股份有限公司
王云燕　远景能源有限公司
吴佳玮　全球能源互联网发展合作组织
（东盟）

Beni SURYADI	东盟能源中心
Akbar Dwi WAHYONO	东盟能源中心
Gabriella IENANTO	东盟能源中心
Adhityo Gilang BHASKORO	东盟能源中心
Rifa FADILLA	东盟能源中心
Annisa Sekar LARASATI	东盟能源中心
Raisha VERNIASTIKA	东盟能源中心
Akbar SWANDARU	东盟能源中心
Zulfikar YURNAIDI	东盟能源中心
Monika MERDEKAWATI	东盟能源中心
Dynta Trishana MUNARDY	东盟能源中心

译　审

魏颖婕　水电水利规划设计总院

翻　译

王艺澄　水电水利规划设计总院
杨雪倩　水电水利规划设计总院

Foreword

China-ASEAN Clean Energy Capacity Building Programme

As the energy sector going through profound changes worldwide, renewable energy has become one of the key areas for global energy development. The world community has agreed on the common mission to accelerate the energy transition towards low-carbon and green development. ASEAN Member States (AMS) have also given a top priority to seizing this historic opportunity of energy reform to convert the clean energy endowment into a new vigor into economic development. Following the philosophy of "innovative, coordinated, green, open, and shared development", China has moved further in upgrading energy production and consumption and optimizing the energy mix and achieved leapfrog development in renewable energy sector. Therefore, it will leverage and build up the mutual strengths and further deepen the cooperation in clean energy development and energy transition towards economic integration of the region for the two sides to jointly launch China-ASEAN Clean Energy Capacity Building Programme.

It is under such context that China and ASEAN have been working together in carrying out China-ASEAN Clean Energy Capacity Building Programme (hereinafter referred to as the "Capacity Building Programme"). Following the vision and framework of the Belt and Road Initiative and based on the dialogue platform of East Asia Summit Clean Energy Forum, the Capacity Building Programme aims to promote clean energy and sustainable development in the region, experience sharing in clean energy policy planning and technology application, and exchanges and capacity building of the core specialists. The Capacity Building Programme targets to build up a hundred ASEAN policy and technical experts in pumped storage, wind power, solar power, nuclear, hydropower, and the combined utilization of multiple energy types in ten years.

The Capacity Building Programme is jointly organized by China Renewable Energy Engineering Institute and ASEAN Centre for Energy.

China Renewable Energy Engineering Institute (CREEI) was established in 1950. Over the years, it has been the only institution dedicated to the administration of hydropower, wind power and solar PV technologies in China. It provides integrated technological support and services in renewable energy policy, resource survey and planning, industry planning, design review, project acceptance, quality supervision, standards compilation, information management, and international cooperation. Commissioned by the National Energy Administration of China (NEA), CREEI oversees the operations of the National Research Center for Hydro and Wind Power, National R&D Center for Hydropower Technologies, Station for Renewable Energy Cost, General Station for Hydropower Project Quality Supervision, General Station for Renewable Energy Project Quality Supervision and National Center for Renewable Energy Information. It is one of the first research and advisory bodies established by NEA.

ASEAN Centre for Energy (ACE) was established in January 1999. As an independent intergovernmental organization representing the interests of the ten ASEAN Member States in energy sector, ACE is committed to promoting economic growth and integration of ASEAN region and initiating and facilitating multilateral collaborations and joint activities in energy sector.

According to the plan of the Capacity Building Programme, CREEI and ACE will work together to organize annually an exchange project targeting at a specialized area (such as pumped storage, wind power, solar PV, nuclear power, hydropower, and the combination of multiple energy types) and invite policy or technical officials from China and ASEAN Member States to gather at seminars. The discussion will focus on policy and technology. The policy side will include the policy framework, comprehensive planning pathways, and industry management tools, while the technology side will cover resource requirements, equipment applications, development and O&M, grid management, risk control, and environment and society.

Under the guidance of NEA, the Capacity Building Programme held two successful exchange projects in 2017 and 2018 respectively with the theme of "Pumped Storage" and "Effective Utilization of Diverse Resources". The capacity building was included in the *China-ASEAN Strategic Partnership Vision* 2030 in 2018.

China-ASEAN Clean Energy Capacity Building Programme 2021 Exchange Project

The third event of China-ASEAN Clean Energy Capacity Building Programme was held in October 2021. The theme was "High Proportion Utilization of Renewable Energy towards a Sustainable Future-Extensive Utilization of Wind and Solar". One of the key topics for discussion during the project was Low Wind Speed Technology (LWST) development and practice. The event was carried online and offline simultaneously. Fifteen ASEAN officials and technical officers from the energy authorities of Indonesia, Laos, Malaysia, the Philippines, and Thailand shared on energy policies. Though not all participants could join the event on the site due to the outbreak of COVID-19, the ASEAN side still showed remarkable enthusiasm with over 90 registered participants and over 1,000 audience watching. The 2021 exchange project served as a high-level platform for profound exchanges. It contributed to the facilitation of ASEAN wind power market development and clean energy cooperation in the region.

Research on Applications of Low Wind Speed Power in ASEAN

In the global efforts against climate change, it has become the common agreement of the countries to reduce carbon dioxide emission. Developing renewable energy and increasing the proportion of clean energy in power supply has risen as an effective means to attain carbon neutrality. The utilization of low wind speed resources, as an important application of wind technology in power generation, is an effective solution to tackle the challenges existing near the load centers, such as low wind speed, scattered distribution of resources, and complex topography. It is a crucial trend in global wind power development. With the progress of wind power technology, the importance of economic feasibility has been growing for project development in low wind speed zones. The tapping of low wind speed resources will significantly expand the potential of wind power development.

Developing renewable energy and reforming the existing fossil-based energy mix is an inevitable choice for ASEAN to address climate change and achieve its emission reduction goals. At the ASEAN Ministers on Energy Meeting in November 2020, the ASEAN Centre for Energy (ACE) officially released the *ASEAN Plan of Action for Energy Cooperation* (*APAEC*) 2016 – 2025 *Phase* Ⅱ: 2021 – 2025. On the basis of the original target to increase the share of renewable energy in the primary energy supply to 23% by 2025, a further target was specified to increase the share of renewable energy in the installed capacity to 35%. To expand the scale renewable energy development will be one of the key missions for ASEAN in its energy and power development.

ASEAN is experiencing slow progress in renewable energy development (excluding hydropower), especially in the exploration of the wind power resources under the traditional requirement on development. According to the relevant statistics, only 1.5% of the land area in ASEAN has an average wind speed above 7 m/s. To support ASEAN in accelerating the development of wind power industry and increase the exploitable wind potential in the region, CREEI and ACE, taking into account both the current global trend and the needs of ASEAN in wind power development, jointly prepared the *Research on Applications of Low Wind Speed Power in ASEAN*. The report was honored to receive support from ACE, Energy authorities of ASEAN Member States, POWERCHINA Zhongnan Engineering Corporation Limited, POWERCHINA Chengdu Engineering Corporation Limited, POWERCHINA Hubei Electric Engineering Co., Ltd., Xinjiang Goldwind Science and Technology Co., Ltd., Zhejiang Windey Co., Ltd., Envision Group, Global Energy Interconnection Development and Cooperation Organization (GEIDCO), and Chinese Renewable Energy Industries Association.

The *Research on Applications of Low Wind Speed Power in ASEAN* is the third title in the China-ASEAN Clean Energy Capacity Building Programme

Technical Materials Publication Series. In the process of preparation, the compilation group took into consideration the development needs of ASEAN and carried out in-depth investigation and analysis on the global low wind speed technology and industry, in a view to provide a useful reference for China-ASEAN low wind speed cooperation and development.

Editor

March 2022

编者的话

中国-东盟清洁能源能力建设计划

当今世界能源形势正在发生深刻的变化，可再生能源已成为国际能源发展的重要领域，加快能源转型，实现绿色低碳发展，已经成为国际社会的共同使命。如何抓住新一轮能源变革的历史机遇，将优越的清洁资源禀赋转化为经济发展的新动力，已经成为东盟国家关注的重点之一。近年来，中国秉持“创新、协调、绿色、开放、共享”五大发展理念，深入推进能源生产消费革命，转变能源发展方式，调整优化能源结构，可再生能源取得飞跃式发展。中国与东盟国家共同开展清洁能源能力建设交流，有助于实现优势互补，进一步深化双方在清洁能源领域的合作，共同促进清洁能源发展和能源转型，推动区域经济一体化进程。

正是在此背景下，中国和东盟共同开展“中国-东盟清洁能源能力建设计划”（以下简称“能力建设计划”）。能力建设计划基于“一带一路”的愿景框架，借力东亚峰会清洁能源论坛的良好对话平台，旨在推动区域清洁能源和可持续发展，分享清洁能源发展政策规划和技术应用等经验，推进相关领域的核心人才交流建设。能力建设计划以“十年百位政策技术骨干”为目标，针对抽水蓄能、风电、太阳能、核电、传统水电、多能互补等专题领域，计划在十年间共同为东盟国家培养百位政策技术骨干。

能力建设计划由水电水利规划设计总院和东盟能源中心共同实施。

水电水利规划设计总院（以下简称“水电总院”）成立于1950年，是中国唯一的水电、风电和光伏发电技术归口管理单位。水电总院为国家可再生能源行业政策研究，资源普查与规划，产业发展规划，工程设计审查、验收、质量监督，技术标准制定，信息管理，以及国际合作等方面提供全方位的技术支持和服务，并受国家能源局委托负责管理国家水能风能研究中心、国家水电技术工程研发中心、可再生能源定额站、水电工程质量监督总站、国家可再生能源信息管理中心等，是国家能源局设立的首批研究咨询基地。

东盟能源中心（ASEAN Centre for Energy，ACE）成立于1999年1月，是独立代表东盟十国能源领域利益的政府间国际组织。ACE致力于推动东盟区域经济发展和区域一体化进程、建立和促进多边合作以及能源领域的协同活动。

根据计划，水电总院和东盟能源中心将每年共同组织一期清洁能源能力建设交流项目，针对一个专题领域（抽水蓄能、风电、太阳能、核电、传统水电、多能互补等），邀请中国和东盟国家的政策或技术官员进行研讨。交流内容分为政策和技术两个方面，政策方面包括政策框架、综合规划思路、产业管理手段等；技术方面包括资源要求、设备应用、开发运维、并网管理、风险控制、环境社会等。

在国家能源局的指导下，能力建设计划已于2017年、2018年分别围绕“抽水蓄能”“多能互补”成功举办了两期培训，并于2018年写入《中国-东盟战略伙伴关系2030年愿景》。

中国-东盟清洁能源能力建设计划2021交流项目

中国-东盟清洁能源能力建设计划的第三期活动于2021年10月举行。本届能力建设主题为“高比例可再生能源助力可持续未来——实现风电和光伏规模化发展”，其中低风速风电技术发展与应用实践是风电领域的重点交流内容。活动以线上、线下相结合的形式开展，来自印度尼西亚、老挝、马来西亚、菲律宾、泰国的15名东盟能源主管部门领导和技术官员进行了政策分享。虽受疫情影响，本届能力建设未能全部在线下开展，但东盟区域的参与热情不减，共有超过90名东盟学员注册报名，超过1000人次参会。2021交流项目为中国和东盟可再生能源领域合作提供了深入交流的高端对话平台，为推动东盟风电市场开发建设，推动区域清洁能源合作起到了积极作用。

东盟国家低风速风电应用潜力研究

在全球应对气候变化的大背景下，减少碳排放已成为各国共识，发展可再生能源、提高清洁电力供应比例成为实现碳中和的有效手段。低风速风电，作为风力发电技术开发应用的重要形式，是解决近负荷区风速低、风资源分散、地形复杂等问题的有效途径，是当前全球风电发展的重要趋势。随着风电技术的不断进步，低风速区域风电开发的经济性不断提高。若能充分挖掘低风速区域风电开发潜力，将大大拓展风电发展空间。

发展可再生能源、改变现有的以化石能源为主的能源结构是东盟国家应对气候变化、实现减排目标的必然之选。2020年11月举行的东盟能源部长会上，东盟能源中心正式发布了《东盟2016—2025年合作行动计划第二阶段：2021—2025》，在原有的到2025年实现可再生能源占一次能源供给比例达到23%的目标基础上，进一步明确可再生能源发电装机容量占比达到35%的目标。扩大可再生能源电力的开发规模将是未来东盟能源电力发展的重要任务之一。

但与此同时，东盟在可再生能源方面（不包括水电），特别是在传统风资源评估标准下的风电发展进程较慢。据有关机构评测，东盟只有1.5%的土地面积的平均风速超过7m/s。为助力东盟加速风电产业发展，拓展区域风力资源利用空间，结合当前全球风电发展趋势和东盟风电发展需求，水电总院和东盟能源中心共同组织编写了《东盟国家低风速风电应用潜力研究》，编写过程中得到了东盟能源中心、东盟10国能源主管部门、中国电建集团中南勘测设计研究院有限公司、中国电建集团成都勘测设计研究院有限公司、湖北省电力勘测设计院有限公司、新疆金风科技股份有限公司、浙江运达风电股份有限公司、远景能源有限公司、全球能源互联网发展合作组织、中国循环经济协会可再生能源专业委员会等的大力支持。

《东盟国家低风速风电应用潜力研究》是“中国-东盟清洁能源能力建设计划”系列技术材料的第三本著作。在其策划和编写的过程中，结合东盟的发展需求，编写组对全球低风速风电技术及产业发展进行了深入调研和系统梳理，力求为中国和东盟低风速风电合作与发展提供有益参考。

编者

2022年3月

内 容 提 要

本书作为“中国-东盟清洁能源能力建设计划”系列技术材料之一，在总结全球风电发展现状的基础上，梳理了低风速风电发展的现状与趋势，分析了低风速风电发展的关键问题。重点以东盟为例，评估了东盟国家的陆上风电资源开发潜力，提出了未来风电发展的重点布局和重点工作，分析了可能存在的问题并提出建议。本书对促进区域低风速风电合作与发展、促进区域能源清洁化发展具有借鉴意义。

本书可供风电技术、风电资源评估与规划、国际工程商务等领域的从业人员阅读，也可供高等院校相关专业师生参考。

Abbreviations

Abbreviated Terms

AC: Alternating Current

ACE: ASEAN Centre for Energy

AMS: ASEAN Member States

ADB: Asian Development Bank

AIIB: Asian Infrastructure Investment Bank

APAEC: ASEAN Plan of Action for Energy Cooperation

APG: ASEAN Power Grid

ASEAN: Association of Southeast Asian Nations

BP: British Petroleum

BCPG: BCPG Public Company Limited

BOI: Board of Investment

BOP: Balance of Plant

CfD: Contracts for Difference

COD: Commercial Operation Date

CREEI: China Renewable Energy Engineering Institute

DC: Direct Current

EGAT: Electricity Generating Authority of Thailand

EPC: Engineering Procurement Construction

EVN: Vietnam Electricity

F+ EPC: Financing+ Engineering Procurement Construction

FiT: Feed-in Tariff

GDP: Gross Domestic Product

GWEC：Global Wind Energy Council
IEA：International Energy Agency
IEC：International Electrotechnical Commission
INDC：Intended Nationally Determined Contributions
IRENA：International Renewable Energy Agency
ITC：Investment Tax Credit
LCOE：Levelized Cost of Electricity
LWSP：Low-wind-speed Power
MODIS：Moderate Resolution Imaging Spectroradiometer
NCC：National Control Center，Thailand
NDC：Nationally Determined Contribution
NDRC：National Development and Reform Commission，China
NEA：National Energy Administration，China
NREL：National Renewable Energy Laboratory，the United States
NREP：National Renewable Energy Program，the Philippines
O&M：Operation and Maintenance
PPA：Power Purchase Agreement
PPP：Public-Private Partnership
PTC：Production Tax Credit
PV：Photovoltaic
R&D：Research and Development
RCC：Regional Control Centers，Thailand
RCEP：Regional Comprehensive Economic Partnership
RE-SSN：Renewable Energy Sub-Sector Network
REC：Renewable Energy Certificate
RMB：Renminbi
RPS：Renewable Portfolio Standard
RPM：Revolutions Per Minute
RUEN：Rencana Umum Energi Nasional，Indonesia
SDG：Sustainable Development Goal
SEB：Sarawak Energy Berhad，Malaysia
SESB：Sabah Electricity Sdn Bhd，Malaysia
TFEC：Total Final Energy Consumption

TNB: Tenaga Nasional Berhad, Malaysia
TPES: Total Primary Energy Supply
TSI: Turbine Supply Installation
UAD: Universal Audio
UAV: Unmanned Aerial Vehicle
UHV: Ultra-high Voltage
USD: United States Dollar
VAT: Value-added Tax
VRE: Variable Renewable Energy
WTGS: Wind Turbine Generator System

Unit
GW: Gigawatt
GWp: Gigawatt peak
kV: Kilovolt
kW: Kilowatt
kWh: Kilowatt-hour
m: Meters
m/s: Meter per second
MW: Megawatt
TWh: Terawatt-hour

Contents

Chapter 3
LWSP Development Conditions and Current Situation in AMS

Chapter 4
Analysis on LWSP Development Landscape in AMS

Chapter 5
Experience and Successful Cases of LWSP Development

Chapter 1

Overview

1.1 Background

(1) Wind power plays an important role in addressing climate change and realizing carbon neutrality. In the global response to climate change, reducing carbon emissions has become a consensus of all countries. Until recently, more than 130 countries in the world have set the goal of realizing carbon neutrality by the middle of this century, six of which have completed the legislation on carbon neutrality. AMS, such as Lao PDR, Myanmar, and Cambodia, are also actively studying the feasibility of realizing carbon neutrality by 2050. As one of the major source of carbon emissions, power generation was responsible for 41% of global carbon emissions in 2019. Developing renewable energy and increasing the share of clean power supply are effective means to achieve carbon neutrality. Wind power, as an important form of renewable energy, has seen steady and rapid growth of installed capacity worldwide in recent years. According to the statistics of International Energy Agency (IEA), the installed capacity of wind power worldwide increased from 3.5% in 2010 to 8.3% in 2019, and was projected to steadily increase up to 14.3% in 2040. Furthermore, with the advancements of wind power technology, there is plenty room to cut down wind power cost while increasing its efficiency.

(2) Developing wind power and other renewable energies help achieve the ASEAN renewable energy targes in installed capacity. Due to the unique geographical location and climatic features, ASEAN region is considered to be the most vulnerable region in the world to the impacts of climate change. At present, fossil energy is still the main source of energy and power in ASEAN. In 2019, its renewable energy accounts for only 13.9% of primary energy supply and about 27.7% of installed power capacity. There is still a significant gap to the target of 23% of renewable energy in the Total Primary Energy Supply (TPES) and 35% in total installed power capacity by 2025 pursuant to the *ASEAN Plan of Action for Energy Cooperation (APAEC) 2016 - 2025 Phase Ⅱ: 2021 - 2025*. Currently, the renewable energy share in ASEAN's power sector has reached 33.4%. More

efforts are needed to keep the momentum of deploying all kinds of renewable energy sources, including wind power, to achieve the target of 35% renewable energy in total installed power capacity by 2025. Moreover, The *6th ASEAN Energy Outlook* (*AEO6*) noted the high potential of solar and wind generation in ASEAN. Developing abundantly of these renewable energy sources, including wind power, will be ASEAN's priority to achieve the APAEC renewable energy target and emission reduction targets.

(3) The low-wind-speed regions in AMS have great potential for development. Wind energy resources have little advantage compared with other renewable energy sources such as hydropower and solar energy in AMS in a tradional sense. Only 1.5% of the land area has the average wind speed of more than 7 m/s at the height of 100 m. However, 13.5% of the land area of AMS is located with low-wind-speed (5 – 7 m/s) resources, so and therefore there is great potential to develop LWSP. At present, most AMS still believe that wind power development requires wind resources with wind speed greater than 7 m/s. Consequently, there is little effort for the AMS to promote wind technology, resulting in the lack of wind power support policies, technical means, and business models in AMS. Until the end of 2019, the installed capacity of wind power accounted for only 0.9% of the total installed capacity, except for a few countries such as Thailand and Vietnam. Most AMS lack of wind turbines. Under the dual pressure of power demand growth and climate change, it is of great significance to fully tap the development potential of LWSP.

(4) The technological breakthrough in LWSP will bring opportunities for AMS to develop wind power. With the continuous improvement of wind power technology, the utilization rate of resources is continuously improved, and the cost of wind power equipment, operation and maintenance decreases gradually, making wind power development more reliable, cost effective, and laying the foundation for developing wind power in low-wind-speed regions. On the other hand, high wind speed regions are almost fully developed, and the global wind power development centres also tend to shift to low-wind-speed regions. Since 2015, the proportion of installed low-speed wind power has steadily increased worldwide. Lower speed wind power accounted for 84% of the new installed capacity in 2019 in China. Technological breakthroughs related to the development of LWSP, especially the success of large-scale LWSP projects in southwest China, which has similar natural geographical conditions with AMS, will provide experience and reference for wind power development in low-wind-speed regions of AMS, and usher in new opportunities for developing wind power in AMS.

1.2 Objectives

The purposes of this research include:

(1) To identify the priorities and key areas of wind power development and to make suggestions on wind power for AMS based on the successful experience of wind power in

the world.

(2) To promote sound development of wind power in AMS by taking stock of resource distribution, analysing features of wind resources, studying the future development of energy and power, and evaluating the development conditions of wind power.

(3) To analyze the layout of wind power development in future and to promote ASEAN's energy transition and ultimately achieve the goal of carbon neutrality.

1.3 Research Thinking and Content

Research framework:

First, analyze the status quo and environment of global wind power development and highlight the development trend and technical features of LWSP.

Second, evaluate and analyse the wind energy resource endowments of AMS, clarify the natural conditions and cost features of wind power development in AMS, make plans on wind power development in AMS, and identify the key areas of wind power development.

Finally, put forward recommendations on LWSP in AMS in line with ASEAN's own features and existing international experience. The framework of research is shown in Figure 1.3.1.

(1) Features of LWSP Development. Review the development history of LWSP, study the development trend of wind power, and analyse the technical features, development models, solutions to environmental and ecological issues, and new technologies adopted for LWSP, taking into consideration of the distribution features of global wind energy resources, and the development status of wind power.

(2) Development conditions of LWSP in AMS. Evaluate wind power resource endowments and technically exploitable wind power resources in AMS with wind speed map, land utilization, and other data, illustrate the state quo of energy and power, renewable energy development policies, and analyse the opportunities and challenges for wind power generation in AMS; analyse the cost, state quo, and trend of wind power development, based on the analysis of cost factors of wind power development and operation and with reference to the existing projects in AMS.

(3) Plans on LWSP generation in AMS. Forecast energy and power development room set medium-term and long-term development goals for wind power based on the overall development of renewable energy in AMS, while taking into account the key locations, regions, and countries for wind power development according to operation scenes of low-wind-speed turbines, resource endowments, comparative advantages, interconnection of power grids, future development, supply and demand of electric power in different countries, and etc.

(4) Suggestions on LWSP development in AMS. Summarize the successful experi-

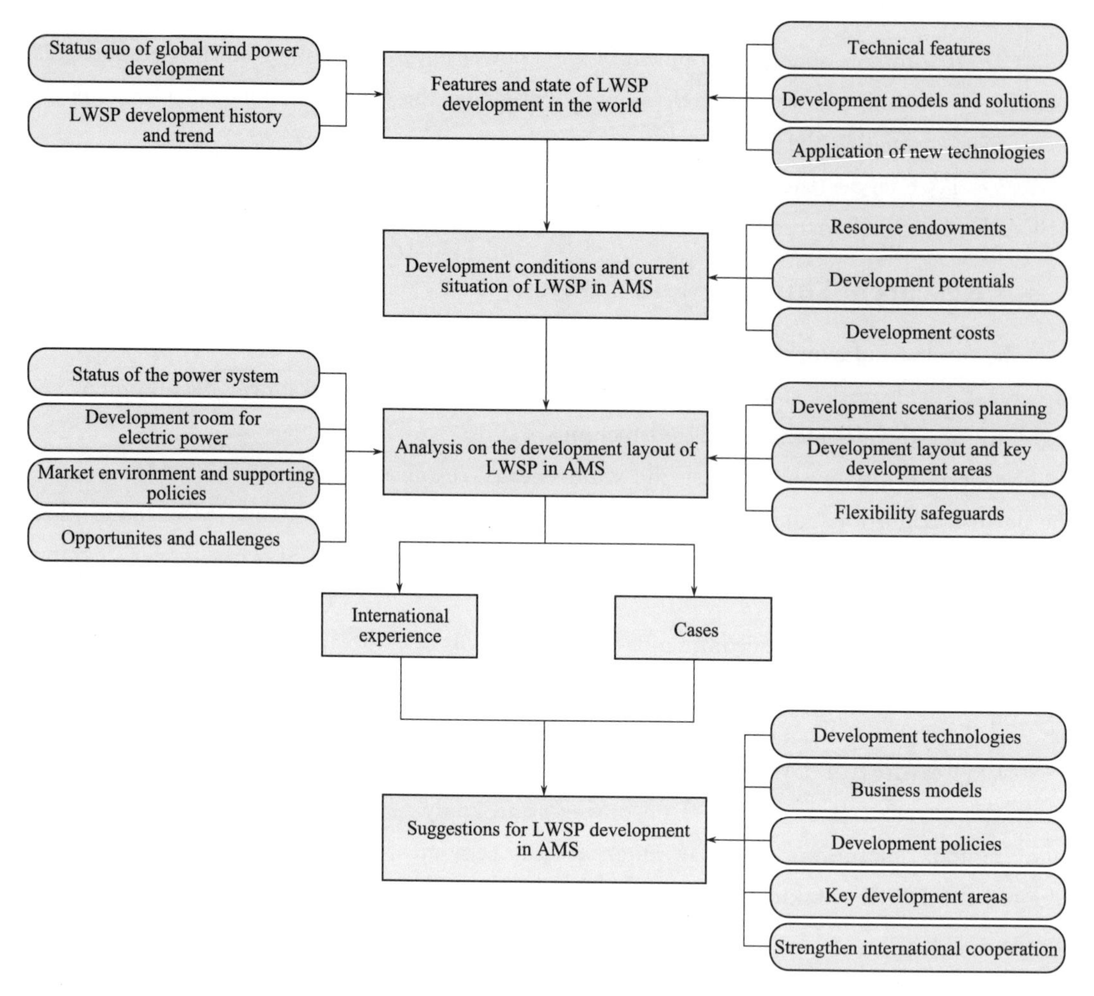

Figure 1.3.1 Framework of the research

ence of LWSP consumption, business model, supporting policies and cost reduction based on specific cases, and make proposals on LWSP development in AMS based on the features of resource distribution, economic and social development, and supporting power grids and equipment in AMS.

Chapter 2

Features of LWSP Development and Current Situation

2.1 Global Wind Power Development

2.1.1 Distribution of global wind resources

Wind is a movement of air caused by the uneven heating of the Earth's surface and atmospheric pressure differences. Wind resources on the Earth's surface are affected not only by the atmospheric pressure under the gradient and geodetic deflection force between different latitudes, but also by the ocean and terrain features. Most of wind resources concentrate in coastal areas and funnel zones of open continents worldwide. It can be seen from the Global Wind Atlas that the regions in the world with better onshore wind resources are mainly in Europe, North America, North Africa, South Africa, Australia, and New Zealand islands. Except for the coastal areas of Central Asia, West Asia, and East Asia, other areas in Asia do not have outstanding wind resources. In China, the regions with better wind resources are mainly in the north (northeast, north, and northwest), and the wind speed in central and southern China is generally lower than 7 m/s. The average wind speed in most regions of AMS is also less than 7 m/s.

2.1.2 Overview of global wind power development

Despite the global economic recession caused by COVID-19 and the decline in energy and electricity demand, wind power continues to grow with the pursuit of carbon neutrality. According to the statistics of Global Wind Energy Council (GWEC), the new installed capacity of wind power in the world exceeded 90 GW in 2020, representing an increase of 52% year on year, and hit a record high. By the end of 2020, the global total installed capacity of wind power reached 743 GW, representing an increase of 14% compared with that in 2019 (see Figure 2.1.1). The installed capacity of onshore wind power was 707 GW, accounting for 95% of the total installed capacity. The rapid growth in global installed capacity of wind power in 2020 was mainly contibuted by the accelerated installment of China and the United States before the preferential policies change and the growth of European market. Africa, South America, the Middle East, and Southeast Asia also performed outstandingly.

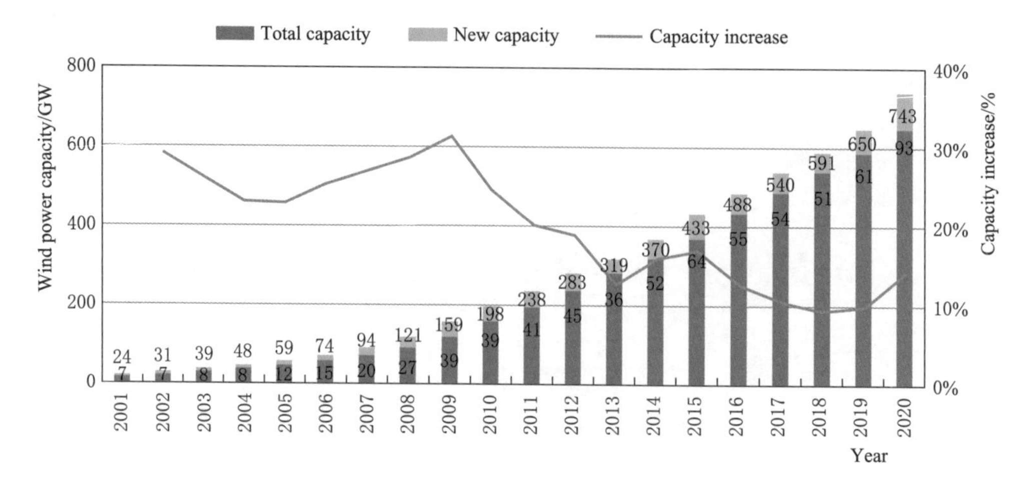

Figure 2. 1. 1 Annual increase and total wind power installed capacities worldwide (Source: GWEC)

Asia, North America, Europe, and South America had the most new wind power capacity in 2020, among which the top five countries were China (52. 0 GW), the United States (16. 2 GW), Brazil (2. 3 GW), Germany (1. 7 GW), and the Netherlands (1. 5 GW) (see Figure 2. 1. 2). Together, these five countries accounted for nearly 80% of new global wind power capacity.

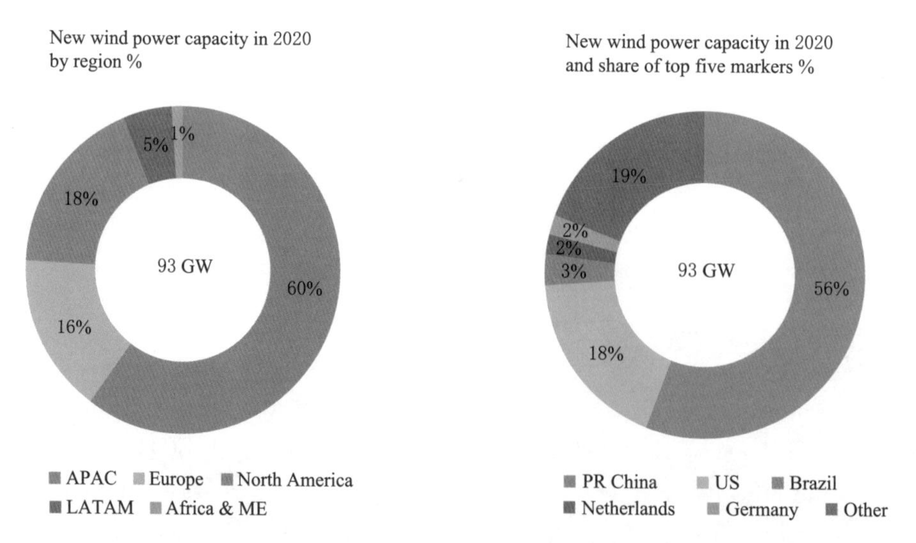

Figure 2. 1. 2 The distribution of new wind power capacity in 2020 (Source: GWEC)

Global wind power output reached 1,262 TWh in 2018, 3. 68 times that of 2010 (342 TWh). The proportion of power generation by wind increased from 8% in 2010 to 19% in 2018.

Based on the findings of British Petroleum (BP), Fitch Solutions, IEA, International

Renewable Energy Agency (IRENA), ASEAN Centre for Energy (ACE), and other organizations, the proportion of wind power output in the total power output in 21 typical countries (wind power penetration) were as shown in Figure 2.1.3. As can be seen from Figure 2.1.3, countries with high wind power penetration were mainly developed countries in Europe and America. Denmark had the highest wind power penetration of 48%, while Ireland, Portugal, Germany, the United Kingdom, and other countries had a penetration of more than 20%. Wind power penetration was about 10% in Morocco, Poland, Brazil, and other countries, representing a medium level. In the United States and China, the wind power penetration was 6.9% and 5.4% respectively, while it was generally low in AMS. The Philippines' wind power penetration exceeded 1.2%, and the penetration was below 1% in Thailand, Lao PDR, and Vietnam, though wind power was developing at a relatively high speed in these countries.

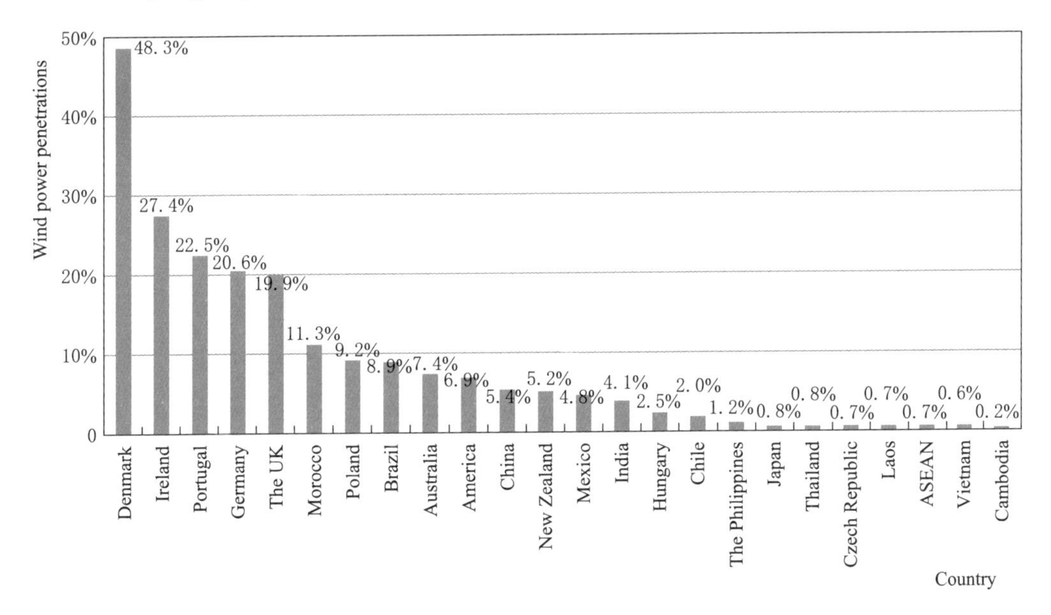

Figure 2.1.3 Comparison chart of wind power penetrations in typical countries in 2019 (Source: BP, Fitch Solutions, IEA, IRENA, and ACE)

2.2 The Development History of LWSP

2.2.1 A new trend for wind power development

With the rapid growth of wind power installed capacity in the past 15 years, wind power projects have been developed around the world. The abundant wind resources is the core factor that promotes the development of wind power projects. At the same time, local tariff policy of the project, consumption on power grid, development of the industrial chain, logistics, transportation, and other factors are also closely related to the development of wind power projects. In the early stage, the government usually ensures the return of project investment by formulating and implementing incentive measures, such as electricity tariff policy to encourage the development of wind power industry. Meanwhile, taking the resources factor into consideration, the global wind power development has been mainly carried out in the regions with better wind resources, and the installment capacity of LWSP is relatively low. After 2015, with the continuous improvement of wind power technology, the wind power development in low-wind-speed regions have been more cost effective, making wind power development in these regions possible. Moreover, the change of tariff policy, the full development of high-wind-speed regions, and other reasons have shifted the focus of wind power development to the middle-wind-speed and low-wind-speed regions. The global development of LWSP entered into acceleration stage.

AMS started to develop wind power relatively late mostly attributed to the limited high-wind-speed resources, though having great potential for low-wind-speed power. As global wind power development gradually shifting to low-wind-speed regions, ASEAN wind power market is expected to become the new driver of global wind power development. A study conducted by Chang and Phoumin, which focuses on modelling and policy implications of wind energy potential in ASEAN, found that the cost to meet electricity demand would increase by 0.7% if no wind energy would be utilized. Thus, wind energy should be promoted in ASEAN to lower costs and reduce carbon emissions.

2.2.2 Development situation

2.2.2.1 Global development

Because wind speeds are lower in low-wind-speed regions, wind turbine generator system (WTGS) with larger swept area per kilowatt are usually needed to capture more wind energy. According to the statistics on LWSP development based on unit selection and data from Wood Mackenzie Power & Renewables, WTGS with swept area per kilowatt above 5 m^2/kW are defined as LWSP projects. In the new wind power projects developed every year with information on turbines recorded, the proportions of LWSP (projects with low-wind-speed WTGS) is shown in Figure 2.2.1, and the structure of global installed LWSP capacity by the end of 2019 is shown in Figure 2.2.2.

As can be seen from Figure 2.2.1, the proportion of LWSP increased from 6.7% in 2015 to 37.3% in 2019 among the new wind power capacity. China led the world in the development speed of LWSP, with the proportion of LWSP in new wind power capacity increasing from 16.7% in 2015 to 84.0% in 2019. By the end of 2019, the total installed low-wind-speed capacity exceeded 40 GW in China, accounting for 72% of the global total. The United States and India had installed capacity of more than 6 GW and 5 GW, respectively, while the other regions had less LWSP developing capacity.

2.2.2.2 Overview of LWSP development in China

Most of the Central, East, and South China have the low-wind-speed regions close to the load centres of the power grid, with the available low-wind-speed resources covering

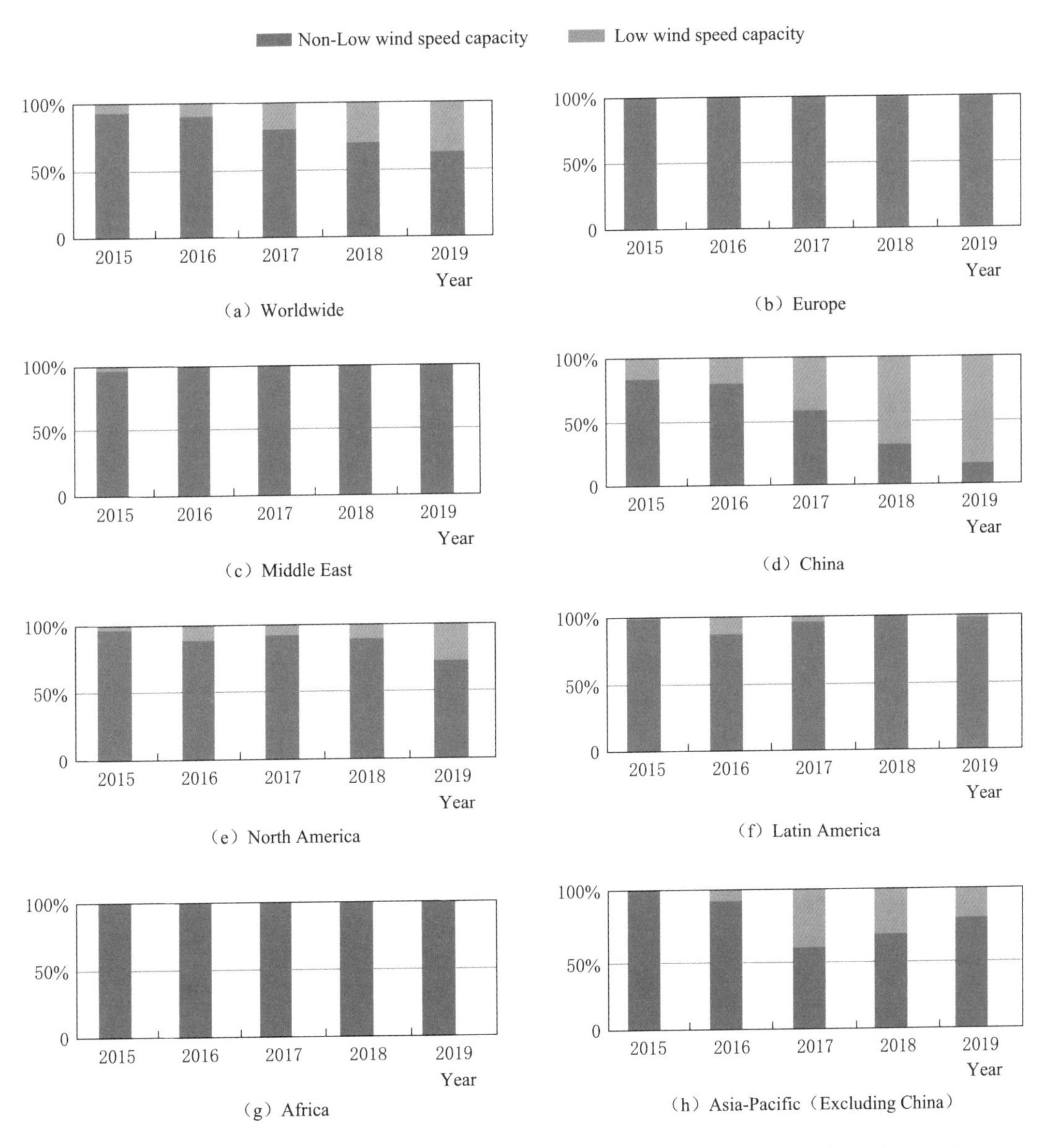

Figure 2. 2. 1 Statistics on the proportion of LWSP installed capacity in newly added wind power capacity in the world and different regions by year (Source: Wood Mackenzie Power & Renewables Database [https: //www. woodmac. com/])

68% of the area with wind energy resources in the country, mainly in Fujian, Guangdong, Guangxi, Anhui, Hunan, Hubei, Jiangxi, Sichuan, Yunnan, and Guizhou. In particular Yunnan, Guizhou, Sichuan, and other southwestern regions, which are located in subtropical monsoon climate areas with complex terrain, relatively large population density, physical, and geographical conditions, are similar to AMS.

China's wind turbine construction was mainly carried out in the northeast, north, and

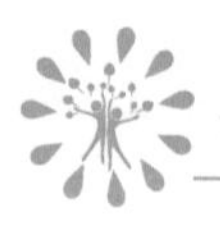

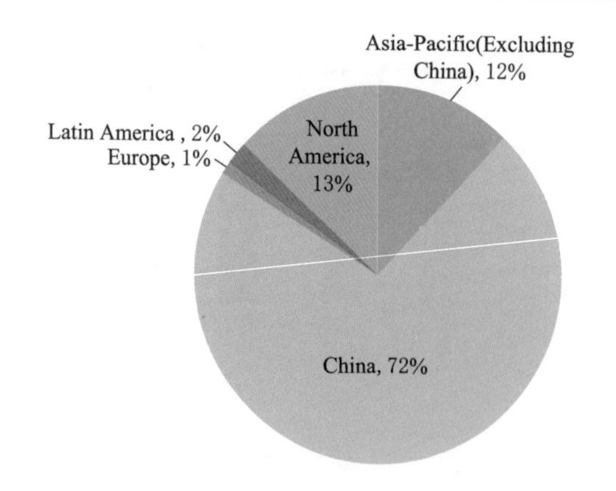

Figure 2. 2. 2 Statistics on the regions/countries with installed capacity of LWSP by the end of 2019 (Source: Wood Mackenzie Power & Renewables)

northwest regions with higher wind speeds during the 12th Five-Year Plan Period. After many years of rapid growth, these regions encountered severe challenges of wind curtailment because they are far away from load centres and lack long distance transmission channels and consumption guarantee mechanisms. The new wind power capacities were also severely affected. Wind power development in China then steadily shifted from the traditional "three Northern regions" with high wind speed to the Central, Eastern, and Southern regions with low onshore wind speed. For the policies, the Chinese government began to encourage wind power investment in Central, Eastern, and Southern regions by adopting regional Feed-in Tariff (FiT) and other policies. Meanwhile, advanced technologies such as large impeller, tall tower cylinder, and intelligent control continued to improve the utilization rate of wind resources over the whole-life cycle management of wind power plants, from the preliminary wind measurement and site selection to the design, construction, operation and maintenance. It also promoted cost reduction and efficiency improvement, and further laid foundation for the development of LWSP.

China has maintained a sound momentum of low-wind-speed power development. In 2020, new wind power capacity increased by 28. 2 GW in Central, East, and South China of low-wind-speed resources, accounting for 39. 4% of the new wind power capacity in China. By the end of 2020, the total installed wind power capacity in Central, East, and South China reached 91. 3 GW, and the proportion of wind power capacity in low-wind-speed regions increased from 12. 8% in 2010 to 32. 4% in 2020 (see Figure 2. 2. 3). Nearly half of the wind power projects with recorded information from 2015 to 2019 adopted low-wind-speed turbines (swept area per kilowatt reaches 5 m^2/kW or above). The installed capacity of low-wind-speed projects reached 33. 1 GW, and 43% of which were carried out in Central, East, and South China.

2. 2. 3 Prospects for LWSP development

As the global energy transition process continues, especially when various countries have formulated their road maps and targets of carbon neutrality, developing LWSP will be one of the priorities in global wind power development, and will help countries with low-wind-speed regions to achieve the renewable energy targets. It is estimated by Wood Mackenzie Power & Renewables that the proportion of LWSP development in the world

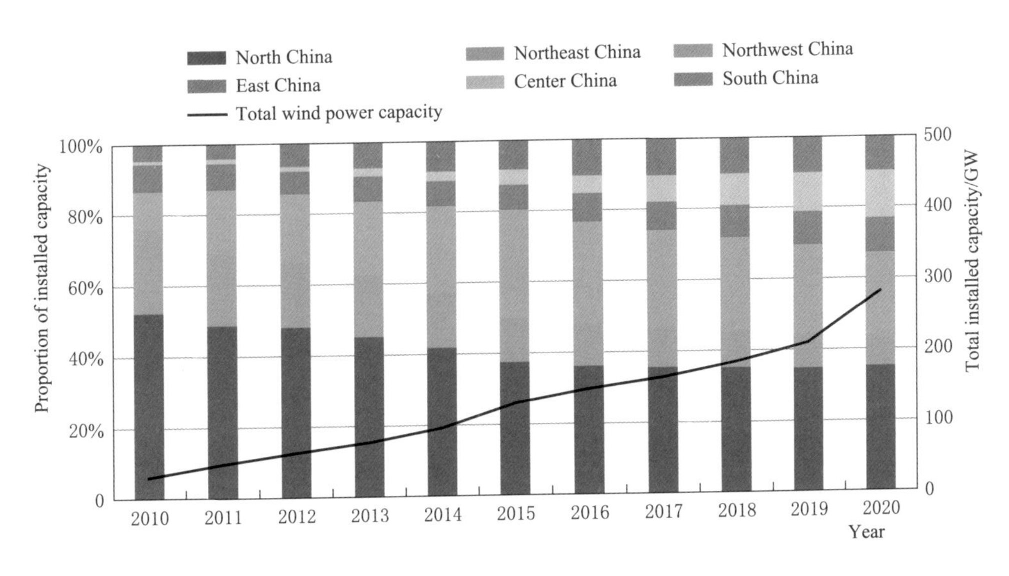

Figure 2. 2. 3 The proportion and total installed capacity of wind power in different regions of China over years (Source: CREEI)

will continue to rise steadily in the next five years. By 2026, LWSP development will account for 28% of wind power installed capacity worldwide (exclusive of China). As the country with the highest installed wind power capacity in the world, China will have LWSP accounting for 95% of the wind power capacity (see Figure 2. 2. 4 and Figure 2. 2. 5).

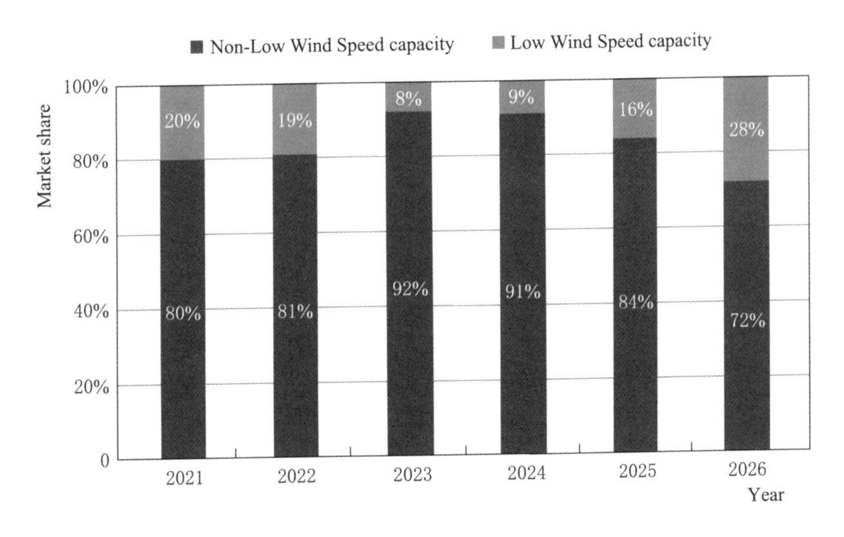

Figure 2. 2. 4 Forecast of global LWSP installed capacity (exclusive of China), 2021 - 2026 (Source: Wood Mackenzie Power & Renewables)

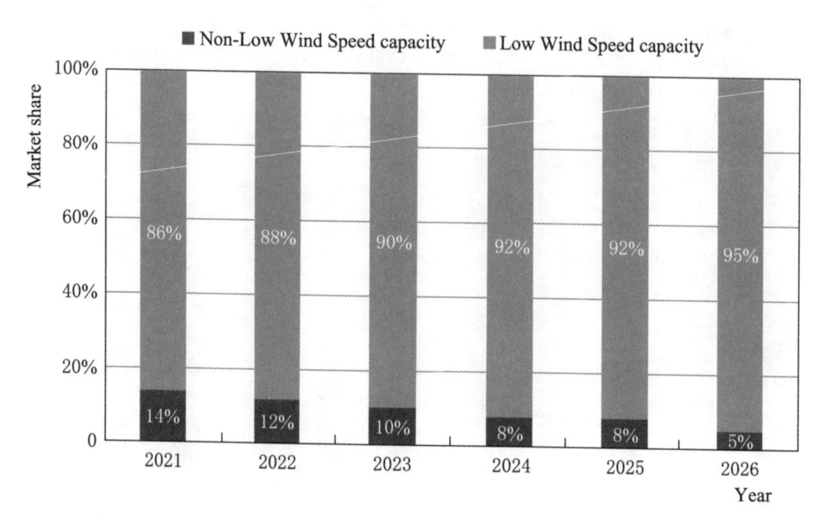

Figure 2. 2. 5 Forecast on the LWSP installed capacity of China, 2021 - 2026 (Source: Wood Mackenzie Power & Renewables)

To adapt to the development of LWSP in the future, the manufacturers of WTGS are shifting to large unit installed capacity, and the industrial chain of rotor blades are also adapting to the trend of large unit installed capacity. According to Wood Mackenzie Power & Renewables, from 2021 to 2026, WTGS with a capacity of 4 MW or above is expected to dominate the market, while turbine blades will be longer, and blades of 70 m or longer will be the mainstream products (see Figure 2. 2. 6 and Figure 2. 2. 7).

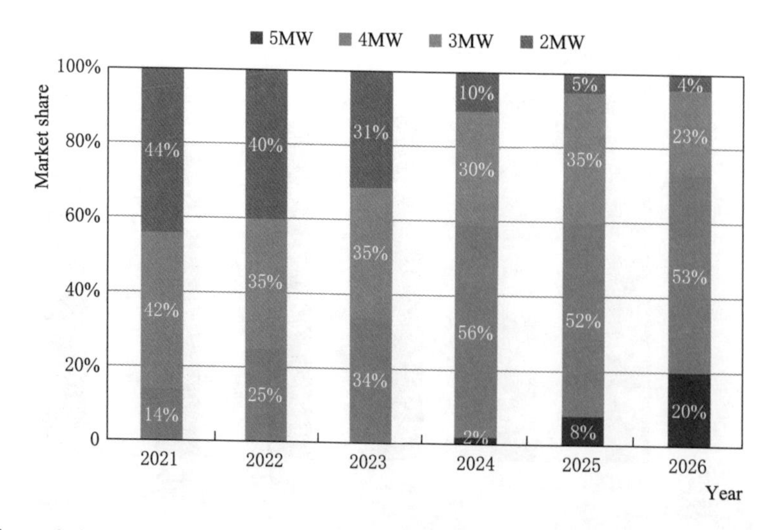

Figure 2. 2. 6 Forecast on the proportion of WTGS with different unit capacities, 2021 - 2026 (Source: Wood Mackenzie Power & Renewables)

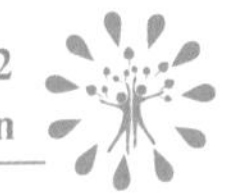

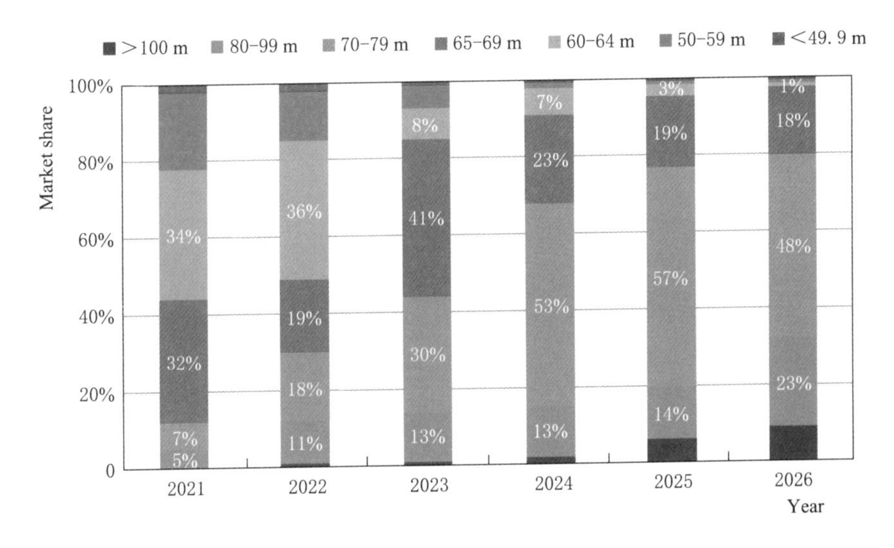

Figure 2. 2. 7 Forecast on the proportion of wind turbines with different blade lengths, 2021 - 2026 (Source: Wood Mackenzie Power & Renewables)

2. 3 Technical Features of LWSP Development

Limited by the conditions, the annual average wind speed and ultimate wind speed are often low, and wind turbines are subject to relatively small wind load in low-wind-speed regions. The current standards on conventional wind farms and requirements on the turbines cannot fully adapt to the features of LWSP projects and WTGS models.

2. 3. 1 Reference standards

At present, the design requirements on wind power system mainly refer to International Electrotechnical Commission (IEC) 61400 - 1. The fourth edition of the standard, published in 2019, stipulates the design, site-specific analysis (site suitability), installation, commissioning, operation, and maintenance of WTGS. IEC 61400 - 1 classified the turbine designs as shown in Table 2. 3. 1. The turbine design is mainly classified pursuant to the annual average wind speed, 10-minute mean reference wind speed (extreme one occurred once every 50 years), turbulence intensity, and other factors. According to these basic parameters and related models, the manufacturers of turbines can carry out load design.

As can be seen from Table 2. 3. 1, the above-shown IEC standard does not specify parameters for low-wind-speed regions. The low-speed wind turbines used in the regions with wind speed slower than 7 m/s is below the lowest class specified in the table (Turbines of Class Ⅲ), and such design is included in Class S in the table. For WTGS applicable to low-speed wind, the manufacturers shall design the load based on the wind resources parameters of the target area to improve the generating capacity of the WTGS and guarantee that the turbine can operate normally when the annual average wind speed is lower than 7 m/s. Consideration shall be given to local situation for the wind power development in low-wind-speed regions. WTGS manufacturers need to have the design capability and experience for low-wind-speed WTGS.

Table 2. 3. 1 Turbine design classification in IEC 61400 - 1

Turbine classification		I	Ⅱ	Ⅲ	S
V_{ave}	m/s	10	8. 5	7. 5	The parameters are determined by the designer according to the specific project
V_{ref}	m/s	50	42. 5	37. 5	
	(m/s) $V_{ref,T}$ in the tropics	57	57	57	
A+	I_{ref} (—)	0. 18			
A	I_{ref} (—)	0. 16			
B	I_{ref} (—)	0. 14			
C	I_{ref} (—)	0. 12			

Note Parameters apply to the height of the turbine hub.
V_{ave} is the annual average wind speed; V_{ref} stands for the 10-minute mean reference wind speed; $V_{ref,T}$ is the 10-minute mean reference wind speed applicable in regions with tropical cyclones; A+ represents a very high turbulence level; A is a high turbulence level; B is a moderate turbulence level; C is a low turbulence level; and I_{ref} is the reference turbulence value.

2. 3. 2 Model selection features of low-speed wind turbines

Limited by the conditions, low-wind-speed regions generally have the features of relatively low annual wind speed and wind energy density. Turbulence changes greatly in some inland regions with complex terrain, and wind shear is usually greater on flat terrain. As a result, there are some distinct traits of turbine design for LWSP compared to the conventional wind power projects.

In terms of wind turbine blades design, low wind speed condition requires a more specific turbine blades design. The use of standard for high wind speed turbine blades will result in less efficient structure. Barnes et al. proposed an improved methodology to specifically design low wind speed turbine blades. The modified low wind speed turbine blades design should be more stiffness-driven to obtain higher efficiency during operation. It is also necessary to develop wind turbines with long blades to capture wind energy for the low wind speed and low wind power density. Usually, the swept area per kilowatt needs to be greater than 5 m^2/kW. For foundation design, customized tower and foundation can be considered for the projects due to the low wind speed and relatively low load requirements to reduce turbine weight and cost. In terms of tower design, the vertical wind shear in low-wind-speed regions is usually larger than that in high-wind-speed regions and the influence of surface roughness. For large wind shears, high towers can be selected to improve wind energy utilization, or weight reduction design can be made. As to turbine

control strategy, low-wind-speed projects are more likely to bring large turbulence, and wind sector management can be considered to avoid large turbulence.

Besides the general requirements on wind turbines in low-wind-speed regions, wind turbines selected should also adapt to local climate and tackle weather disasters because of the climatic features of AMS. AMS tend to have relatively high annual relative humidity. Thus, it is necessary to take environmental control measures such as adding dehumidifiers and upgrading the anti-corrosion of WTGS. There is more rain and more frequent thunderstorms in these regions, so it is necessary to provide lightning monitoring equipment. For coastal projects affected by tropical cyclones, it is necessary to evaluate the extreme wind speed according to the provisions in the appendix of IEC 61400 - 1: 2019 and select appropriate models. Meanwhile, standby power supply devices and typhoon warning mechanisms should also be added.

By the end of 2020 with the wind power capacity accounting in China's low-wind-speed regions for 32.4% of the total installed capacity, the country accumulated a lot of experience in low-speed wind turbine development and customization. LWSP projects launched in Yunnan, Guizhou, and Guangxi have natural conditions similar to AMS, in particular, and can provide experience for AMS to develop LWSP projects.

2.3.3 Wind farm development process

Wind farm development mainly involves screening the areas suitable for wind farm construction with technical judgment on wind resources, and conducting compliance procedures in accordance with the relevant national policies, and procedures for the screened areas to obtain the investment and construction permit of government. Wind farm development will go through the following process: project site selection and preliminary preparation, project compliance preparation, Power Purchase Agreement (PPA) application, project approval, and construction permit (Figure 2.3.1). The project site selection and preliminary preparation involve macro-site selection, wind measurement, micro-site selection and economic evaluation of the investment, project approval, and project company registration. Project compliance preparation involves land acquisition (including woodland and logging permit), evaluation on mineral, cultural relics, military zones, environmental assessment, and etc. PPA is the electricity purchase and sale agreement signed after negotiation with the buyer and usually involves the grid-connection license. Project approval and construction permit refer to the development and construction approval of the target project given by relevant government departments.

There is not much difference between LWSP farms and the conventional wind farms in terms of development process, but LWSP regions are usually close to the load centres. In the process of macro and micro-site selection, more attention should be paid to minimizing the negative environmental impact and avoiding disturbing nearby residents.

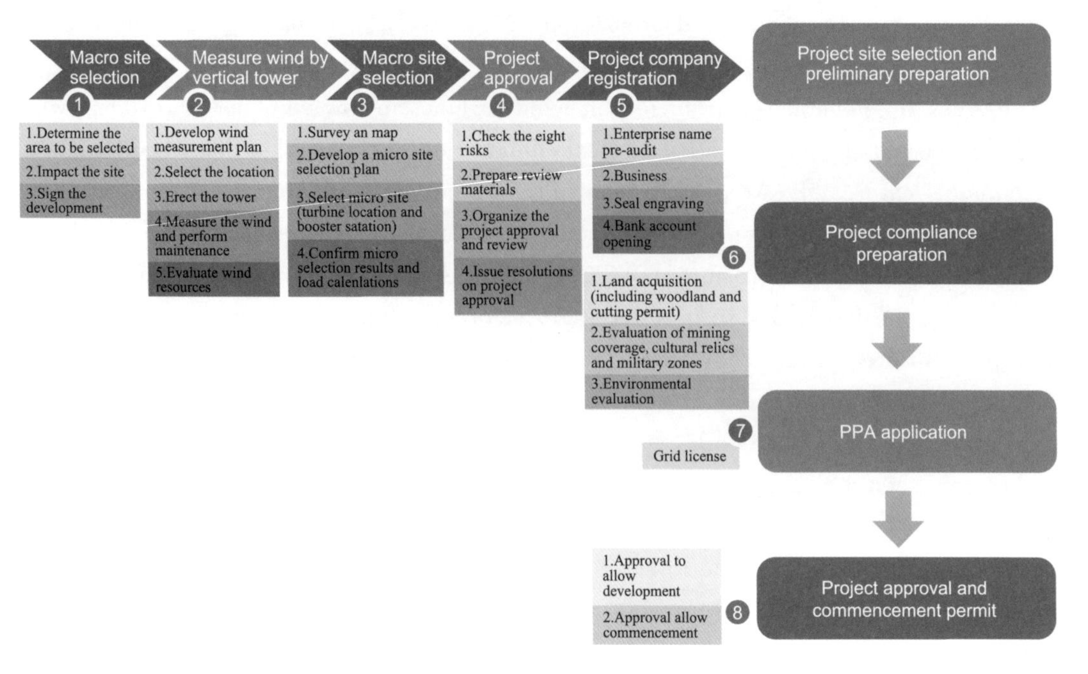

Figure 2. 3. 1 Wind farm development process

2. 4 Development Models and Solutions

LWSP installations are usually close to load centres, with low-wind-speed regions having relatively dense population, diverse terrains, complex wind conditions, and higher requirements on the development technology. Therefore, it is necessary to choose diversified development models according to the natural conditions, and find out the technical solutions for the wind power development under specific scenarios.

2. 4. 1 Development models

2. 4. 1. 1 Centralized development

Centralized wind power projects have all WTGS output of the wind farm transmitting to a substation before being connected to the transmission grid. The scale of centralized wind power generation is generally above 50,000 kW, and the investment cost can be further reduced through large-scale development, which enables uniform large-scale long-distance transmission and management of electricity. In general, new step-up stations and transmission lines are needed, and centralized management centres need to be set up for the operation and maintenance.

This mode mainly applies to regions with relatively sparse population and developed engineering technology systems, otherwise it is difficult to effectively reduce the development cost.

Due to the need for connection to the transmission grid, centralized wind power projects

are planned by the governments. There are two main ways for enterprises to obtain the development and management right in wind power projects: upon concession and approval. Resources are allocated generally by agreement or competition.

2.4.1.2 Distributed development

Decentralized wind power generation project refers to the projects which are located near the load centre and transmit the electricity to the grid nearby. The electricity is consumed locally and does not require large-scale long-distance power transmission. The electricity is transmitted to the power distribution network nearby, consumed and regulated in the power distribution system.

The distributed development mode may hardly apply to relatively densely populated regions. With distributed WTGS, electricity is carried to and consumed by the distribution network nearby, with less land used and less impact on the ecological environment. The decentralized wind power development is a good supplement to centralized wind power. The large-scale development of distributed wind power in densely populated regions is common around the world. In Germany, for example, wind power is mainly developed in small scale and a decentralized mode, and 98% wind farms have no more than five wind turbines. Under this mode, the installed wind power capacity in Germany is still as high as 60 GW, and the installed capacity per square kilometre of land area reaches 175 kW.

In combination with the diverse demands of urban and rural construction and eco-tourism development, decentralized wind power projects can also be developed alongside power plants, industrial parks, docks and ports, beautiful villages with scenes, and special towns for comprehensive development and utilization, in a view to meet the needs of electricity consumption, realize the economic and intensive land use, and optimize and beautify the ecological environment.

2.4.1.3 Community wind power

Community wind power is similar to distributed wind power. Denmark, Germany, and other countries are promoting the development of distributed wind power by policies, and wind power is very closely related with communities and residents. Community residents can benefit from wind power investment by jointly purchasing shares of wind farms, which also significantly improves public acceptance of wind power. As a pioneer in community wind power, Denmark has 80% distributed wind farms as community wind power. Danish municipal energy agencies not only purchase community wind power, but also participate in the investment, and play a very important role in the popularization of community wind power in Denmark. In Germany, more than 75% of distributed wind power can be classified as community wind power. The owners of community wind power include local farmers, independent companies, and cooperatives. Independent companies buy public equity stakes in community wind power farms, and energy companies hold more and more shares of the community wind power farms.

2.4.1.4 Smart mini-grids

As a small, decentralized, independent system of power generation and distribution, smart mini-grids are composed of distributed power supply, electrical load, power distribution facilities, monitoring and protection devices, and etc. They can basically achieve the internal balance with electric power and energy. Mini-grids are powered by a combination of two or more sources of energy such as wind, solar energy, natural gas, and biomass, and usually have the function of energy storage. At present, smart mini-grids have become an important distributed power development mode in the United States, Europe, and Japan.

Due to the wide distribution of low-wind-speed resources, the development and utilization of wind resources well adapt to local conditions and can bring good economic benefits. Therefore, LWSP is expected to become one of the important components of smart mini-grids. Wind power has played an important role in several smart mini-grid applications such as the independent power supply systems used for supporting the local mining industry on Dinagat Island in the Philippines, and the small island power system of the Sta. Ana project in Vietnam.

2.4.2 Application solutions for different development environment

2.4.2.1 Technical solutions for typhoon

The AMS encounter more tropical cyclones and some of which could be typhoons in general. From the perspective of wind power development, typhoon is highly destructive and has complex wind regimes and other features. During a typhoon, the wind direction changes abruptly; the turbulence is abnormal; the extreme wind speed goes beyond the design load of the turbines; and the load acting on the turbine exceeds the design load several times, and thus destroying the turbines. To avoid the risk brought by the complex wind regime, the turbines are usually shut down before the typhoon. But the wind speed before and after a typhoon is very high, so a lot of power can be generated. In most coastal areas, wind farms are operated at full load only under the impact of a typhoon. Therefore, the economic efficiency and reliability of the turbines should be considered comprehensively.

Similar to AMS, China's southeast coast is also regularly hit by typhoons. To minimize the damage caused by typhoons to wind power plants, China has made standards and specification for the design and installation of wind turbines, and formulated solutions for wind power development. Through these projects, China accumulated rich experience in wind power development in the coastal areas and realized the large-scale development of wind power in the conditions of typhoons. Several wind farms survived violent typhoons safely.

2.4.2.2 Solutions for mountainous areas

Some regions with relatively good wind resources, such as Central and South Vietnam, and North the Philippines, also have high altitudes and mountainous terrain, and it is difficult to construct wind power projects. One of the main construction problems for

LWSP farms is the transportation of blades in mountains. The transport vehicles special for blades as shown in Figure 2. 4. 1 can realize blade lifting, rotating, and horizontal movement by hydraulic steering of the rear wheel and turning strickles. With small turning radius, less pavement occupied, reversing convenience, and other features, the vehicles can keep away from obstacles around the road in the process of transportation, reduce road construction and reconstruction, shorten the construction period, cut capital investment, and minimize the damage in the mountains.

Figure 2. 4. 1 Transport vehicles special for LWSP blades in mountains

2. 4. 2. 3 Solutions to ecological and environmental issues

Noise, harm to flying animals, visual pollution, and other ecological and environmental problems may emerge in the construction and operation of wind farms. Especially for LWSP farms which are closer to the load centres, it is more necessary to take certain technical measures to reduce the adverse effects of these ecological and environmental problems.

(1) Noise. Noise is a common problem in LWSP development. It usually needs to be solved in the following ways:

1) Work out the correct influence range of noise in site selection and arrangement by improving the accuracy of modelling and optimizing the algorithm;

2) Properly select the site and optimize the number and location of wind turbines in the wind farms according to the results;

3) Optimize the turbine design by setting serrated trailing edges on the blades (Figure 2. 4. 2), adopting low-noise air foils, damping the inside of the shells, and other measures to reduce the noise from the source in the turbine design;

4) Reduce the noise by reasonably optimizing the control parameters, reducing the Revolutions Per Minute (RPM), adjusting the initial Pitch Angle (Pitch) or changing the propeller in advance, and etc. ;

5) Separate noise reduction control can also be carried out for a certain period of time in terms of control optimization.

Figure 2.4.2 Serrated trailing edges

(2) Effects on birds and bats. To reduce the impact of wind turbines on flying animals such as birds and bats, the following factors should be considered in the construction and operation of wind power projects:

1) Proper site selection (keeping away from bird habitats and protected areas such as wetlands and ridge edges, as well as migration bottlenecks and flight paths for endangered species). Research has shown that wind turbines located more than 1,600 m (about 1 mile) from bird habitats do not have a significant impact on bird activity;

2) Dynamic and intelligent control. Referring to the atlas and real-time bird migration forecasts issued by bird protection departments, wind turbines can be temporarily stopped or slowed down during the peak hours of bird flight;

3) Increase the cut-in wind speed of turbines. Birds tend to avoid flying at windy night, and the risk of collision can be avoided to some extent by increasing the cut-in wind speed;

4) Drive away bats by using sound waves. Bats use sound to navigate—echolocation. Therefore, ultrasonic silencers or Universal Audio (UAD) devices can be installed on wind turbines to interfere with the hearing systems of bats nearby to drive them away and avoid collisions;

5) Wind turbine appearance design. Irradiate wind turbines with ultraviolet light or paint them purple to reduce their attraction to insects such as flies and moths, and to avoid indirect attraction to birds and bats (Figure 2.4.3).

(3) Visual solutions. The shadow cast by and the light reflection from the rotating wind turbine blades may make people feel dizzy and upset, and thus affect their normal work and life (Figure 2.4.4). The solutions including:

1) Adjusting the size or location of wind turbines when selecting the turbines to mini-

mize potential impacts;

2) Shut down the problematic wind turbine when the shadow flicker effect may occur;

3) Paint them by using non-reflective coating to avoid reflection of light and shadow.

Figure 2. 4. 3 Wind turbines set on a pier near the city centre of Liverpool, England, painted to reduce the impact on wildlife

Figure 2. 4. 4 Impacts of light and shadow from wind turbines

2. 5 New Technologies for LWSP Development

With the continuous change of the global wind power industry chain and technology, there are more and more innovations and breakthroughs in LWSP technology. At the same time, bold innovations are made by drawing on the successful experience of the Internet, big data, Internet of Things, and artificial intelligence. LWSP projects are more profitable and have tremendous opportunities.

2. 5. 1 New technologies to improve design efficiency and quality

(1) Unmanned Aerial Vehicle (UAV) survey. Compared with conventional wind power projects, LWSP projects make use of scattered wind resources, and it is more important to get detailed information such as topography and landform on the site during the preliminary design. Usually, both macro-site and micro-site selection can be time-consuming, laborious, and have high risk. Since the field of view is limited, it is impossible to grasp the whole picture upon site survey, and the turbine locations often need to be retrofitted before and after the survey, making it inefficient. UAV technology provides solution to such problem (Figure 2. 5. 1). In the early stage of planning, site selection, survey, and design, corresponding 3D models can be created by using the high-definition pictures taken by UAVs to have the panorama of the wind farm with a wide aerial view, pro-

viding accurate data for designers, save manpower, and improve efficiency.

Figure 2.5.1 UAV survey

(2) New wind measurement technologies and schemes. Reliable wind measurement data serve as a key factor in the whole life-cycle management of wind farms and can provide a more reliable basis for the decision-making at wind farm planning stage, the promotion of generating capacity, and the operation and maintenance management. More sensitive to wind speed, LWSP projects should ensure a more comprehensive and reliable wind measurement. Based on conventional ways, measuring wind by laser radar, virtual anemometer tower, and intelligent wind measurement management platforms can help to get the whole picture of wind resources on the sites, to avoid highly risky wind power development areas and to maximize the project benefit.

(3) Digital wind farm site selection planning and design platform. Compared with medium and high-speed wind farms, low-speed wind farms have higher requirements on site selection, wind resource assessment and wind farm design, and require more reliable technology and supporting means. Covering micro-site selection, customized selection, engineering design and other activities, digital wind farm site selection planning and design platform can produce wind resources simulation results, engineering design results, economic evaluation estimates and other results faster, better, and less expensive. It can solve the problems that determine the success of the project, such as the inaccuracy of wind atlas, the long wind resource simulation cycles, and the low profit caused by the uncertainty of site selection, and guarantee the maximum rate of return.

2.5.2 New technologies to reduce the difficulty and cost of construction

Due to the complexity of wind resources, wind energy and resources may vary greatly between different projects, and even different turbine locations in the same project; wind regimes and environmental conditions of the locations may go beyond the existing design standards. At the same time, the wind turbine with long blades and high tower cylinders

required by LWSP projects also greatly increases the equipment manufacturing, transportation, and hoisting costs. These problems can be solved by customized turbine location design and installation scheme.

2.5.2.1 Customized turbines for different locations

The environmental adaptability of the wind farm can be improved by replacing the traditional "universal wind farm design" with "customized design for turbine locations", considering more configuration options in the turbine power, wind wheel diameter, tower height and other dimensions, and customizing the control strategies.

Design of the most suitable tower for the site need to base on the wind resources and tower costs, using advanced simulation model of load computer system, for example, increasing generating capacity by raising the height of towers to meet the requirement on turbulent flow resistance in low-wind-speed, hilly and mountainous regions, and areas with a high maximum shear stress value, and adopting the split tower cylinder to reduce the load, manufacturing cost, and transportation difficulty brought by the high tower (Figure 2.5.2).

Segmented blades can lower the requirements on production, equipment, and tooling and the difficulty of manufacturing, transportation, and installation of super-long blades on site (Figure 2.5.3). Customized design can ensure the operational stability of turbines at each location and the return on investment in low-speed wind farms.

Figure 2.5.2 Split high tower cylinder

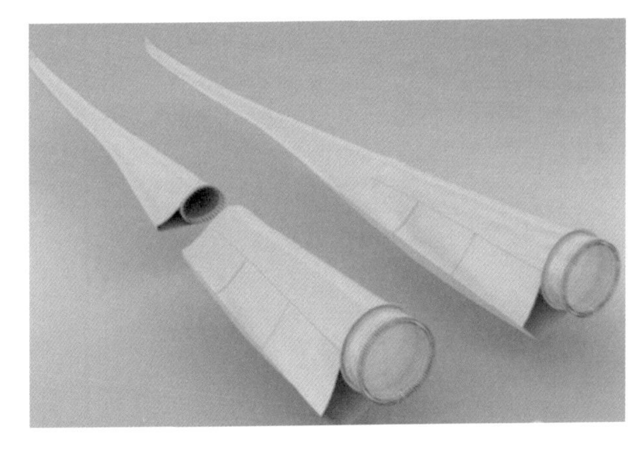

Figure 2.5.3 Segmented blades

2.5.2.2 Customized installation schemes and equipment

Generally, WTGS of low-wind-speed projects are arranged at the top in relatively high-altitude mountains, so the turbine hoisting platform is small, and the wind farm is in harsh hoisting environment. Therefore, the site levelling and land expropriation cost more; the installation is inefficient; and the construction period is long. Customized installation schemes and equipment can effectively solve the problem, cut the cost, and reduce the environmental damage caused by turbine infrastructure construction.

The elephant leg tooling scheme is a customized auxiliary installation equipment that is small in size, easy to transport, and suitable for turbine installation in the location with a small area, thus can greatly improve the installation efficiency and reduce the hoisting cost (Figure 2.5.4).

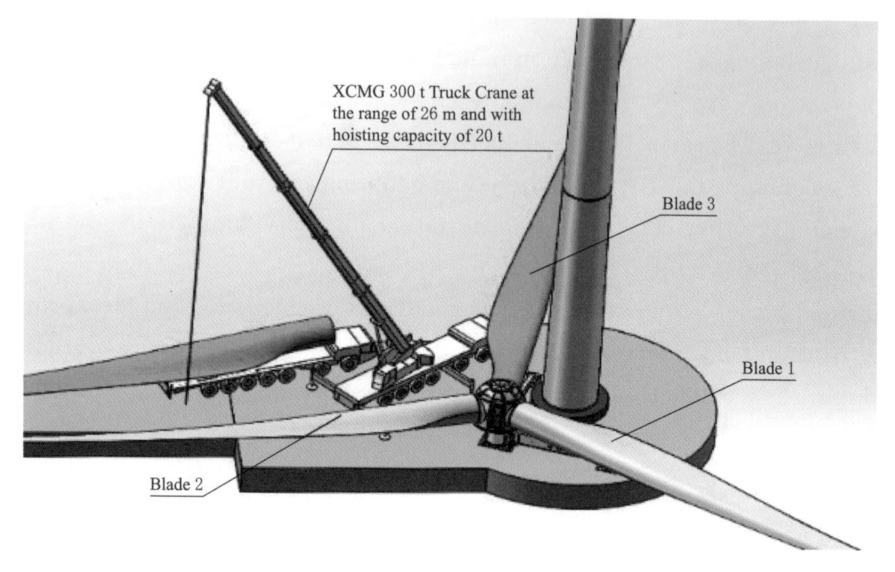

Figure 2.5.4 Elephant leg tooling scheme

Figure 2.5.5 Single-blade installation scheme

The single-blade installation adopts special installation tools (Figure 2.5.5). The special tools can reduce the land area and crane tonnage used and are more suitable for wind power projects in jungle areas, projects of high hoisting cost or WTGS larger than 3 MW.

Self-climbing crane can better solve the problems traditional large-scale installation equipment has (high cost, large size, easy to be affected by gale and other inclement weather) that result in slow transition and maintenance inefficiency. It can greatly reduce the operative surface required for turbine installation, and effectively cut the installation cost of LWSP projects (Figure 2. 5. 6).

Figure 2. 5. 6　Self-climbing crane

2. 5. 3　New technologies to improve the operation efficiency of wind farms

(1) Cluster optimization. As a set of wind farm control system, the cluster optimization system provides a solution platform for intelligent full-life-cycle diagnosis, analysis, and optimization of the turbine performance with the help of big data, machine learning, and cloud platform, under the coordinated control at the wind farm level. Capable of self-learning, the cluster optimization system can optimize the operation management over each turbine, fully tap the potential of each location, and find the optimal combination of performance and cost based on the condition of wind turbines. Cluster optimization system can automatically collect and issue parameters, and exchange data with turbines and the monitoring system (Figure 2. 5. 7).

Figure 2. 5. 7　Cluster optimization technology

(2) Unit-power raising. As one of the effective means to improve the capacity of single generator in low-wind-speed projects, unit-power raising technology is used to optimize the blade shape and power curve, and improve the aerodynamic efficiency of the blades according to the specific operation. The practice shows that unit-power raising technology can improve the power generation efficiency of turbines by 2.5%-7% based on theoretical generating capacity without affecting the certification and load of the turbines.

(3) Laser radar control. Multi-dimensional information of the incoming air flow can be perceived in real time and accurately captured by the laser radar control technology in advance. Combined with the advanced control algorithm, the technology can move with the wind and follow the trend to significantly reduce the unit load, optimize the power, and effectively improve the generating capacity of LWSP projects.

(4) Operation, maintenance, and inspection by UAV. The relatively harsh natural conditions of low-wind-speed projects in mountainous areas make the daily operation, maintenance, and inspection more difficult. The UAV has positioning accuracy, is not affected by the complex terrain of mountains, flies according to the planned or pre-determined route, and clearly reflects the external damage of generators and blades as well as the failure of electrical circuits and other problems by taking high-precision pictures or videos. The use of UAV for blade and electrical circuit inspection greatly lowers the risk of manual inspection, improves inspection efficiency, and reduces operation and maintenance costs (Figure 2.5.8).

Figure 2.5.8 Operation, maintenance, and inspection by UAV

Technological progress is a strong driving force for the development of LWSP and market investment. With the continuous technological innovation and change, wind turbines will be more efficient and intelligent in the future, and the full-life-cycle cost and efficiency of wind power projects will be further optimized from wind measurement at the early stage to equipment selection and then to operation and maintenance. New business models, application scenarios, and market demands for LWSP will emerge with more powerful products, and high-quality and efficient services, therefore further driving the development of wind power in ASEAN region and contributing to global carbon neutrality.

Chapter 3

LWSP Development Conditions and Current Situation in AMS

3.1 Geography

ASEAN consists of 10 countries located in Southeast Asia region, namely Brunei Darussalam, Cambodia, Indonesia, Lao PDR, Malaysia, Myanmar, the Philippines, Singapore, Thailand, and Vietnam. ASEAN faces the Pacific Ocean to the east and the Indian Ocean to the west. It is at the "crossroad" between Asia and Oceania, and between the Pacific Ocean and the Indian Oceans. It is at 92°10′E - 141°05′ E and 28°32′N - 11°15′S, with a total area of about 4.436 million km^2. Among AMS, Vietnam, Lao PDR, Cambodia, Myanmar, Thailand, and Malaysia are located in Indo-China Peninsula in Southeast Asia, and the other countries in Malay Archipelago.

The northern part of the Indo-China Peninsula is connected to China. The terrain is high in the north and low in the south. Mountains and rivers extend from north to south, forming the terrain of mountains and rivers distribution alternately and vertically. Most rivers in the Peninsula originate from Southwest China, with large drops and rapid currents at upper reaches, and wider courses and slower flow and sediment deposits downstream. Thus, the Indo-China Peninsula countries, such as Lao PDR, Myanmar, and Cambodia have rich hydropower resources, while Thailand and Vietnam in flood plains and delta areas are important agricultural production areas in ASEAN. There are more than 20,000 islands in the Malay Archipelago, most of which have rugged terrain, numerous mountains, volcanoes, and earthquakes; the plains are narrow and small and mostly distributed in coastal areas, such as Indonesia, Malaysia, Brunei Darussalam, and the Philippines. Among them, Indonesia is the largest country by area in ASEAN and the largest archipelagic country in the world with 17,508 islands.

ASEAN is close to the equator, with tropical rainforest climate from 10 degrees north latitude to 10 degrees south latitude, tropical monsoon climate prevailing from 10 degrees

to 20 degrees north latitude, and a small area of alpine plateau climate in the northern part of Indo-China Peninsula. Dominated by tropical monsoon climate, Indo-China Peninsula has a dry season and a rainy season in one year. The rainy season has more rainfall and high humidity, and the dry season has sufficient high temperature and sunshine. Dominated by tropical rainforest climate, Malay Archipelago is hot and rainy all year round, with dense tropical rainforests.

Affected by geographical location and climate, the ASEAN region is very rich in solar energy resources. For the most countries and areas in the region, the total annual solar radiation is more than 1,750 kWh/m^2, and the technical exploitability of solar power can reach 9,929 GW, but wind energy resources with more than 7 m/s are not abundant.

3.2 Endowment of Wind Energy Resources

3.2.1 Resource characteristics

The wind energy resources in the 10 AMS are quite different. According to the data provided by Global Wind Atlas, wind speeds are below 7 m/s in most areas of the AMS. The countries with relatively good wind resources are mainly in the central and southern Indo-China Peninsula and some coastal areas of Malay Archipelago. The areas with certain wind power development potential are mainly in Lao PDR, Thailand, Vietnam, south-central Myanmar, Cambodia, and the Philippines.

To quantitatively describe the region's wind energy resources, an overview of wind energy resources in AMS is shown in Table 3.2.1 according to the classification of wind energy resources in NB/T 31147—2018 Technical Code for Wind Energy Resource Measurement and Assessment of Wind Power Projects.

Table 3.2.1 Overview of onshore wind energy resources in AMS

No.	Country	Wind Energy Resources
1	Brunei Darussalam	The multi-year average wind speed is in the range of 1 - 3 m/s, the average wind power is less than 50 kW/m^2, and the wind energy resources in most areas are lower than D-1 level.
2	Malaysia	The multi-year average wind speed is in the range of 1 - 3 m/s, the average wind power is less than 50 kW/m^2, and the wind energy resources in most areas are lower than D-1 level.
3	Indonesia	The multi-year average wind speed range of most areas is 1 - 3 m/s, the multi-year average wind speed range of a small part of coastal areas is 3.5 - 5 m/s, and the average wind power range is 50 - 150 kW/m^2. The wind energy resources in most areas are lower than D-1 level.

Continued

No.	Country	Wind Energy Resources
4	Singapore	The multi-year average wind speed ranges from 1 – 3 m/s in most areas, from 3.5 – 5 m/s in a small part of coastal areas, and from 5.5 – 7 m/s in a very small part of coastal areas. The average wind power is in the range of 50 – 150 kW/m², and the wind energy resources in most areas are lower than D – 1 level.
5	The Philippines	Wind energy resources are mainly concentrated in the central, southern, and partly northern areas, but the northernmost part is severely affected by typhoons. The multi-year average wind speed ranges from 5 – 7 m/s in the central and southern areas, and from 6 – 8 m/s in a small part of the northern area. The average wind power is in the range of 150 – 500 kW/m². The wind energy resources are Level 2 in most areas and Level 3 – 4 in a small part.
6	Thailand	The distribution of wind energy resources is relatively concentrated, and the wind energy resources in the southern and central areas are better than those in the northern areas. The multi-year average wind speed ranges from 5.5 – 8 m/s in the central areas, and from 6 – 8 m/s in a small part of the northern area. The average wind power is in the range of 200 – 500 kW/m², and the wind energy resources in most areas are Level 2.
7	Myanmar	The distribution of wind energy resources is relatively concentrated, and the wind energy resources in the southern and central areas are better than those in the northern areas. The multi-year average wind speed ranges from 5 – 7 m/s in the central and southern regions. The average wind power is in the range of 150 – 400 kW/m², and the wind energy resources in most areas are Level 2.
8	Lao PDR	The distribution of wind energy resources is scattered. The northern, central, and parts of southern areas have good wind energy resources while the southern area has concentrated resources with a multi-year average wind speed ranging from 5 – 7 m/s. The average wind power is in the range of 150 – 400 kW/m², and the wind energy resources in most areas are Level 2 – 3.
9	Vietnam	The distribution of wind energy resources is scattered. There are excellent wind resources in all parts of Vietnam. The resources in the southern coastal areas are relatively concentrated, and the multi-year average wind speed ranges from 5 – 8 m/s. The average wind power is in the range of 150 – 500 kW/m², and the wind energy resources in most areas are Level 3.
10	Cambodia	The distribution of wind energy resources is concentrated, and the wind energy resources in the southwest and northwest areas are better than those in the central and northeast areas. The multi-year average wind speed ranges from 5 – 6.5 m/s in the southwestern and north-western areas. The average wind power density is in the range of 150 – 300 kW/m², and the wind energy resources in most areas are Level 2.

Box 3.1 The classification for wind energy resources in NB/T 31147—2018 Technical Code for Wind Energy Resource Measurement and Assessment of Wind Power Projects.

NB/T 31147—2018 Technical Code for Wind Energy Resource Measurement and Assessment of Wind Power Projects is approved by the National Energy Administration (NEA) of China and is prepared by China Renewable Energy Engineering Institute (CREEI) to guide the wind energy resource measurement and evaluation of onshore wind farm projects. It covers basic regulations, wind energy resource measurement, and wind energy resource assessment. The Code classifies wind energy resources in different areas according to the characteristics of regional wind resources such as wind power density and wind speed at a height of 100 m. The classification is shown in Table 3.2.2.

Table 3.2.2 Classification of wind power density at a height of 100 m

Levels of wind energy resources	Wind power density/(W/m^2)	Wind speed/(m/s)
D-1	<150	5
D-2	150-190	5.4
D-3	190-230	5.8
1	230-280	6.2
2	280-410	7.1
3	410-540	7.7
4	540-670	8.3
5	670-800	8.8
6	800-1,070	9.7
7	1,070-2,570	13

The classification of wind resources levels is helpful for formulating corresponding wind power development plans and policies according to the characteristics of regional wind resources, and optimizing the utilization of wind resources.

3.2.2 Development conditions

In addition to the characteristics of wind energy resources, the development of wind power projects is also related to the conditions, such as terrain, ground cover, transportation, a power grid to be connected, and destructive wind speed.

In terms of the terrain, the flatter the terrain and more concentrated the wind energy resources distribution is, the lower development cost and higher power generation income

will be for equal conditions of resources. For the plain areas, the terrains are flat and are desirable for wind power development, such as in central and southern Cambodia, southern Lao PDR, southern and central Thailand, southern Vietnam, southern Myanmar, and northern and central the Philippines. For the inland high-altitude mountainous areas, wind energy resources are concentrated and also have good conditions for development, such as in northern Myanmar, northern Cambodia, northern Thailand, and northern Lao PDR.

In terms of ground cover, the areas with a low ground roughness are suitable for wind power project development in general, such as bare land with sparse or low vegetation, or grassland. Forest land with abundant ground cover and high ground roughness is unfavourable for constructing wind power projects. However due to the evolving to high hubs and large blades of LWSP, the increase in ground roughness caused by vegetation will have little adverse impact on the development of wind power projects.

In terms of transportation, the developed transportation facilities and the extensive road networks will improve the development and construction conditions of large-scale wind power bases, facilitate the entry and transportation of construction equipment and materials, and reduce the development costs of the wind power bases. According to the statistics of Natural Earth, the total mileage of ASEAN trunk roads is about 131,000 km, of which the length of major highways exceeds 6,800 km, and the length of roads is about 24,000 km. In the countries with rich wind energy resources, such as the central and southern Thailand and southern Vietnam, transportation conditions are convenient.

In terms of power grid for connection, the better the condition of the grid infrastructure is, the lower the cost of connecting to the grid from a large-scale wind power base will be which is conducive to the development of centralized wind power generation. The densely distributed areas with 115 kV transmission lines are mainly in the areas with rich wind energy resources, such as the whole of Thailand, central Lao PDR, and the coast of Vietnam. Most regions with relatively rich wind energy resources also boast favourable grid infrastructure, which is conducive to the development of wind power projects.

In terms of destructive wind speeds, the coastal areas are affected by typhoons which are destructive to wind turbines, especially the east the Philippines which is frequently hit by typhoons. Indonesia, Malaysia, and Brunei Darussalam in the south of ASEAN are located on both sides of the equator and affected by the equatorial calm zone, so the wind is small throughout the year and not disturbed by typhoons. Therefore, typhoons will attenuate rapidly after landing and have little impact on inland wind farms, except for the Philippines.

3.3 Potential of Wind Power Development

3.3.1 Technically exploitable capacity

3.3.1.1 Research methods

The wind power exploitable capacity is evaluated at three levels: theoretical exploitable capacity, technically exploitable capacity, and economic exploitable capacity, as shown in Figure 3.3.1.

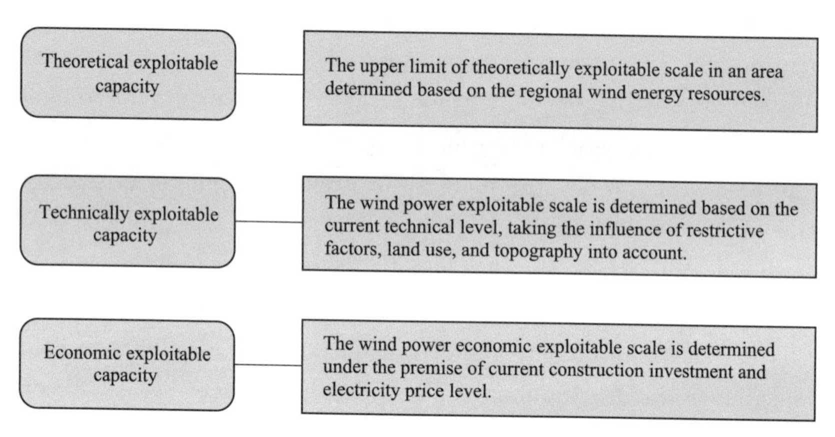

Figure 3.3.1 Evaluation of wind power exploitable capacity

This research mainly focuses on the evaluation of technically exploitable capacity of wind power. The common method for evaluating the technically exploitable capacity is based on researching the regional wind energy resource map and land use situation, eliminating the unusable area restricted by topography, altitude, land use and wind speed, and obtaining a suitable region for development of wind power (i.e., wind power technologically developable area). Then, the installed capacity per unit area of the region is determined based on the regional wind energy resource conditions, and finally the wind power installed potential of the technologically developable region is obtained. In the ASEAN, only the northern part of Myanmar and the eastern part of Indonesia have higher altitudes. At the same time, wind power technology for mountainous areas has matured, so this research does not take mountainous terrain as an eliminated factor in wind energy resource evaluation.

Wind energy resource evaluation steps are shown in Figure 3.3.2.

3.3.1.2 Data sources

The data required for the research covers wind energy resource maps, land use data, and protected area data. The wind energy resource maps come from Global Wind Atlas, and the land use data comes from Moderate Resolution Imaging Spectroradiometer (MODIS). With unique climate conditions, AMS have high vegetation coverage. The land use is mainly woodland, grassland, and shrubs. Irrigated farmland is mainly distributed in

southern Vietnam, southern Myanmar, and estuary deltas and plains in central and southern Thailand.

Step 1: Generate a map of ASEAN wind energy resources	
Source: Global Wind Atlas	Wind energy resource data: average wind speed at a height of 100 m average wind power density at a height of 100 m
Step 2: Eliminate the areas with poor wind energy resources from the perspective of resource amount	
Required data: Average wind speed at a height of 100 m	Elimination condition: average wind speed <5 m/s
Step 3: Eliminate unusable areas from the perspective of land use and protected areas	
Required data: land use, protected areas	Unusable areas: nature reserves, woodlands, artificial ground surfaces, towns, water bodies, etc. Eliminate unusable areas and determine technically developable areas
Step 4: Estimate the technically exploitable capacity of wind power in combination with land use situations	
Required data: technologically developable area, land use	1. Set land use coefficients according to ground surface features Priority use area (utilization coefficient=1): bare land, land with sparse vegetation, grassland Partially usable area (utilization coefficient <1): farmland, shrub land, cultivated land/vegetation mosaic 2. Determine key parameters Installed capacity per unit area: 10 MW/km^2 3. Calculation of technically exploitable capacity Technically exploitable capacity of wind power=region area×installed capacity per unit area×utilization coefficient
Step 5: Calculate the technically exploitable capacity based on different wind speeds	
Required data: average wind speed at a height of 100 m	Range: 5－6 m/s, 6－7 m/s, more than 7 m/s Calculate the technically exploitable capacity of wind power based on different wind speeds

Figure 3.3.2 Flow chart for calculation of technically exploitable capacity of wind power

The protected area data comes from Protected Planet. Thanks to excellent natural conditions, almost all AMS have many natural reserves, including the central and southern Indo-China Peninsula, and the central and northern regions of the Philippines with relatively good wind energy resources.

3.3.1.3 Analysis of the results of technically exploitable capacity

According to the conditions shown in Figure 3.3.2, the technologically developable regions are mainly concentrated in the central and southern Indo-China Peninsula, central and northern the Philippines, and some island coastal areas of Indonesia. Among them, Vietnam, Thailand, the Philippines, Lao PDR, and Myanmar have relatively large areas

for wind power development technologically.

The technically exploitable wind power capacity in ASEAN is counted by regions and shown in Table 3. 3. 1 and Figure 3. 3. 3, and the spatial distribution is shown in Figure 3. 3. 4. The technically exploitable wind power capacity in AMS is 1,112. 30 GW, of which Thailand, Vietnam, the Philippines, Myanmar, and Indonesia are among the top five.

Table 3. 3. 1 Statistics of technically exploitable capacity of wind power and mean annual utilization at different wind speeds

Country	Technically exploitable capacity/GW				Mean annual utilization hours/h
	More than 7 m/s	6 - 7 m/s	5 - 6 m/s	Total	
Brunei Darussalam	0. 00	0. 00	0. 02	0. 02	2,000
Cambodia	0. 44	1. 82	47. 51	49. 76	2,152
Indonesia	4. 75	20. 42	80. 36	105. 53	2,255
Lao PDR	6. 07	32. 17	33. 23	71. 47	2,360
Malaysia	0. 02	0. 14	1. 28	1. 44	2,223
Myanmar	2. 64	11. 58	141. 08	155. 30	2,244
The Philippines	36. 56	51. 62	98. 01	186. 19	2,472
Singapore	0. 00	0. 00	0. 01	0. 01	2,000
Thailand	3. 13	43. 84	294. 43	341. 40	2,190
Vietnam	24. 03	39. 01	138. 13	201. 17	2,346
Total	77. 63	200. 60	834. 07	1,112. 30	2,313

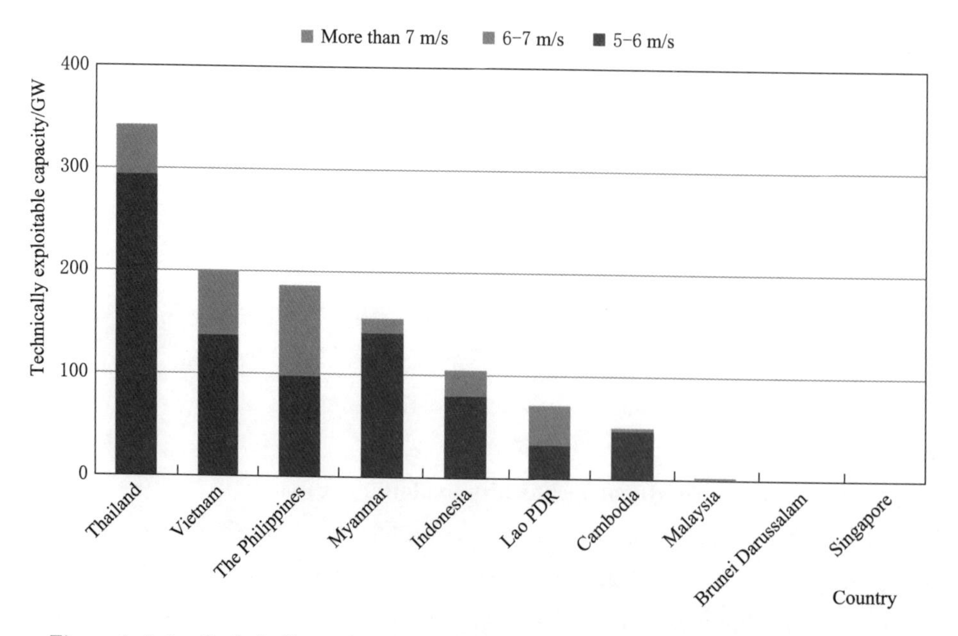

Figure 3. 3. 3 Technically exploitable capacity of wind power in ASEAN (by country)

Within the technologically developable areas of the AMS, most areas are in the lower wind speed range of 5 – 6 m/s, and there is great potential for low-speed wind power development. The technically exploitable capacity of 5 – 6 m/s wind speed range is 834.07 GW, accounting for 75% of the total; that of 6 – 7 m/s is 200.60 GW, accounting for about 18%; while that of more than 7 m/s is only 77.63 GW, mainly distributed in the Philippines and Vietnam. From the perspective of the distribution of technically exploitable capacity between 5 – 6 m/s, Thailand has the largest capacity, 294.43 GW, followed by Myanmar with 141.08 GW, and Vietnam with 138.13 GW. In terms of annual utilization hours, within the technically exploitable areas of the AMS, the mean annual utilization hours is 2,313 h. The mean annual utilization hours of the Philippines, Vietnam, and Lao PDR exceed 2,300 h, exceeding the average level of AMS. For Singapore, Brunei Darussalam, and Malaysia with low wind energy potential, there is limited commercial potential for wind power projects development.

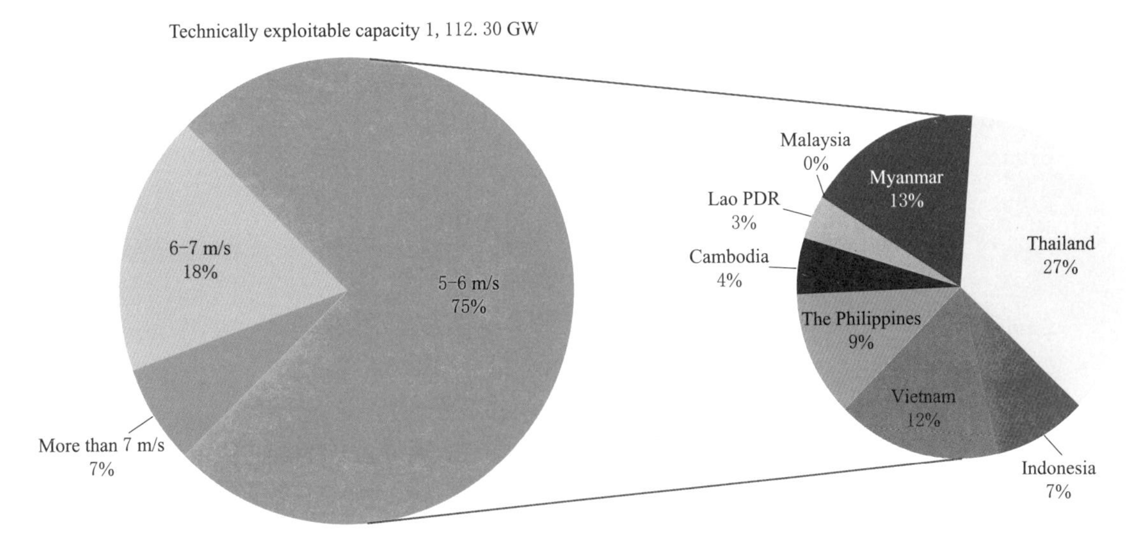

Figure 3.3.4 Proportion of technically exploitable capacity at different wind speeds in ASEAN region, and proportion of capacity in 5 – 6 m/s range by AMS

3.3.2 Development situation

In recent years, wind power in ASEAN region has grown steadily, and the growth is mainly due to the increase of wind power installed capacity in Thailand, the Philippines, and Vietnam. However, the overall scale of wind power development is still relatively low. As of the end of 2019, the total exploitable wind power capacity in ASEAN region was 2,344 MW, accounting for only 0.3% of the total installed capacity of wind power in the world. Some countries with a certain development potential for low-speed wind power such as Lao PDR, Myanmar, and Cambodia, have not yet started the development of wind power (Figure 3.3.5).

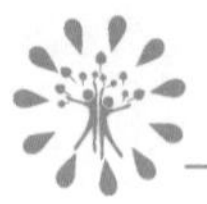

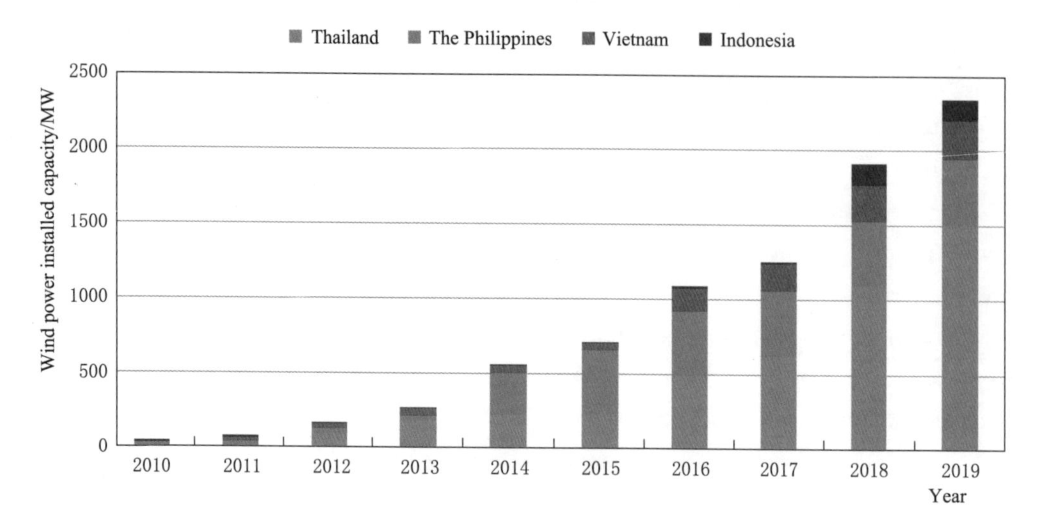

Figure 3. 3. 5 Wind power installed capacity in AMS, 2010 to 2019
(Source: ASEAN Energy Database System, AEDS)

Brunei Darussalam

Brunei Darussalam's technically exploitable capacity of wind power is 0. 02 GW. The wind power potential is low. By the end of 2020, the installed capacity of wind power was zero.

Cambodia

Cambodia's technically exploitable capacity of wind power is 49. 76 GW. In the past 10 years, Cambodia's electricity consumption has increased dramatically. Since 2010, the average annual growth rate of electricity demand has grown to 20%, and the electricity supply can barely meet the demand. Until the end of 2019, Cambodia had not yet started the development of wind power projects, but the country will have great development potential in the future.

Indonesia

Indonesia's technically exploitable capacity of wind power is 105. 53 GW. Before 2018, the installed capacity of wind power was only 2 MW. In 2018, the newly installed capacity of wind power was 142 MW, but no new wind power projects were put into operation after 2018. In consideration of energy transition, Indonesia's wind power development will usher in new opportunities.

Lao PDR

Lao PDR's technically exploitable capacity of wind power is 71. 47 GW. Until the end of 2019, the development of wind power projects has not yet started. At present, the power constitution is dominated by hydropower, and power supply is in serious shortage in

dry season. The development of a wind-solar hybrid system will effectively improve the seasonal imbalance of power output and bring a new opportunity for wind power development.

Malaysia

Limited by wind resource conditions, Malaysia's technically exploitable capacity of wind power is only 1.44 GW. Until the end of 2019, the wind power installed capacity was zero.

Myanmar

Myanmar's technically exploitable capacity of wind power is 155.30 GW, ranking fourth among the 10 AMS, and has a great potential for wind power development. Until the end of 2019, the installed capacity of wind power was zero. The power system had problems such as power supply lag and high cost of electricity. With the demand for electricity continuing to grow, wind power will become an important source for Myanmar's renewable energy if construction cost can be reduced.

The Philippines

The Philippine's technically exploitable capacity of wind power is 186.19 GW, 19.6% of which is in the wind speed range above 7 m/s. With the largest proportion of the technically exploitable capacity being in the wind speed range above 7 m/s, it has a huge potential for wind power development. In 2014 and 2015, the installed capacity of wind power grew rapidly, but the increase of wind power capacity slowed down after 2015. Until the end of 2019, the installed capacity of wind power was 442 MW.

Singapore

Singapore's technically exploitable capacity of wind power is 0.01 GW, and the commercial potential of wind power is low. Until the end of 2019, the installed capacity was zero.

Thailand

Thailand has the largest technically exploitable capacity for wind power in ASEAN region, reaching to 341.40 GW. The wind power development started early and maintained a steady growth. Until the end of 2019, the installed capacity of wind power reached 1,507 MW, making the country with the largest installed capacity of wind power in ASEAN region.

Vietnam

Vietnam's technically exploitable capacity of wind power is 201.17 GW, and there is a potential to develop wind energy resources. In recent years, the installed capacity of wind power has increased steadily. Until the end of 2019, the total installed capacity of wind power reached 251 MW. There is still a great potential for wind power development in the future.

3.4 Wind Power Development Costs and Trends

3.4.1 Current development costs

3.4.1.1 Construction cost

1. Division of construction cost

From the perspective of wind farm projects, the construction cost mainly includes WTGS and ancillary facilities, logistics, hoisting, site electrical facilities (including set-up substation) and installation, civil works, and transmission works.

Generally, the construction of wind farm projects can be divided into three parts: procurement and installation of WTGS, on-site electrical equipment installation and civil works, and transmission works. The procurement and installation of WTGS is called Turbine Supply Installation (TSI). The main scope of work includes WTGS and ancillary equipment supply, logistics, and hoisting. On-site electrical and civil works is called Balance of Plant (BOP), which mainly includes on-site collection lines and foundation construction of WTGS. Some investors may incorporate the construction and installation of set-up substations into the scope of work. The transmission works usually include the transmission line from the wind farm to the connection point of power system and the transformation of the substation. In the TSI part, the purchase expense of wind turbines accounts for 30% to 50% of the total cost. Therefore, the reduction of wind turbine cost can effectively lower the construction cost of wind power projects.

2. Current situation of construction cost

From the perspective of the main component of construction cost (wind turbine cost), the price of wind turbine prices has shown a significant downward trend in the world (Figure 3.4.1). Among them, the prices of Chinese wind turbines and Vestas wind turbines have fallen by more than 40% from 2008 to 2019. According to the latest research on the price of wind turbines, first-class manufacturers such as Vestas wind turbines have lowered the price to about 500 USD/kW, and Goldwind, one of the Chinese wind turbine producers, has lowered the price to 430 USD/kW.

The main reason for the decrease of wind turbine cost is the advancement of wind turbine technology. Generally, the units with large installed capacity are conducive to reducing cost per kilowatt. According to IRENA, compared with 2010, the rotor diameter and the unit installed capacity of the wind turbine units have increased by 40% on average in 2018 for the major countries with a certain scale of wind power installed capacity, such as Sweden, Germany, China, and Canada. In addition, wind turbine manufacturers will also customize the capacity and rotor blade length of wind turbines according to customer requirements. The average rotor diameter and wind turbine capacity of wind farms in the ASEAN region have also increased significantly. At present, the unit capacity of bid-winning onshore wind turbines is generally above 3 MW.

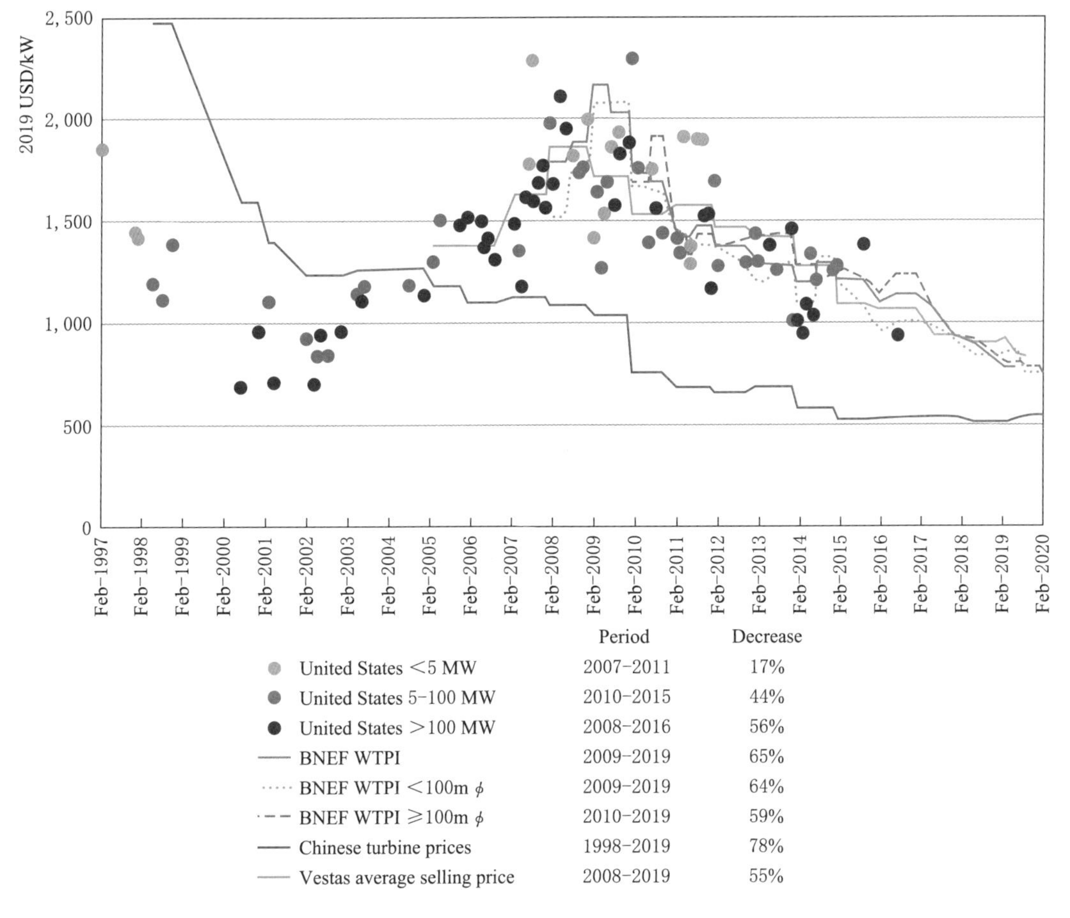

Figure 3.4.1 Change trend of wind turbine cost, 1998 to 2019 (Source: IRENA)

From the perspective of overall construction cost, the average construction cost of global wind power projects in 2019 was 1,473 USD/kW, which was 24% lower than that in 2010; among them, the construction cost of wind power projects in 2019 in Asian (except China and India) was 2,368 USD/kW, 5.3% lower than that in 2010. This discrepancy on cost per kilowatt between AMS and global average is mainly due to the advancement of wind turbine technology particularly in developed countries such as Europe and United States, which also are advanced in the industrial chain, logistics, transportation, and other factors affecting the cost of wind power development.

In addition to the ex-factory price of wind turbines, the factors that have a greater impact on the construction cost include logistics and transportation conditions, local policies and land use restrictions, and labour cost. Therefore, construction cost may vary greatly in different projects.

According to the prices of the completed wind power projects in ASEAN collected in this research, the cost of wind power construction in Thailand is 1,820 - 1,930 USD/kW, with an average of 1,870 USD/kW; while the cost of wind power construction in Vietnam is 1,580 - 2,250 USD/kW, with an average of 1,920 USD/kW. In general, the development of wind power in ASEAN is still in its infancy, with limited logistics and transportation conditions and lack of incentive policies in some countries. Therefore, the construction cost of wind power projects in the AMS at present is generally slightly higher than the international average.

3.4.1.2 Levelized Cost of Electricity

The Levelized Cost of Electricity (LCOE) of wind power averages the cost of the full lifecycle of a wind power project to the unit power generation. The calculation parameters involve the project - and economy - related parameters, such as operation period, construction cost, operation and maintenance cost expected annual energy production, inflation rate, capital cost, and tax rate.

In this research, the calculation of LCOE for wind power refers to the method mentioned in the report developed by National Renewable Energy Laboratory (NREL) and ACE. The cost rate of projects completed in AMS that combines the research situation and the calculation parameters of the LCOE has been determined, considering the construction cost, operation and maintenance cost, loan interest rate, loan ratio, tax rate, and inflation rate. The calculated average LCOE is shown in Figure 3.4.2.

Box 3.2 Calculation method of wind farm LCOE

The LCOE calculation formula:

$$LCOE = \frac{INVT - \sum_{n=1}^{N} \frac{Dep^{n}}{(1+DR)^{n}} TR - \frac{RV}{(1+DR)^{n}} + \sum_{n=1}^{N} \frac{AC^{n}}{(1+DR)^{n}} (1-TR)}{\sum_{n=1}^{N} \frac{IEP}{(1+DR)^{n}}}$$

where: $INVT$ is the initial investment, that is the dynamic investment or construction cost of a project, consisting of Engineering Procurement Construction (EPC) cost and construction period interest; Dep is the present value of tax deductions due to depreciation in the full life cycle; RV is the present value of the residual value of fixed assets (the construction cost that is not consumed during the entire life cycle); AC is the present value of the project's operating costs in the full life cycle, and the operating cost including operation and maintenance cost and financial cost; IEP is the expected average annual energy production; DR is the discount rate; TR is the income tax rate; N is the operating period.

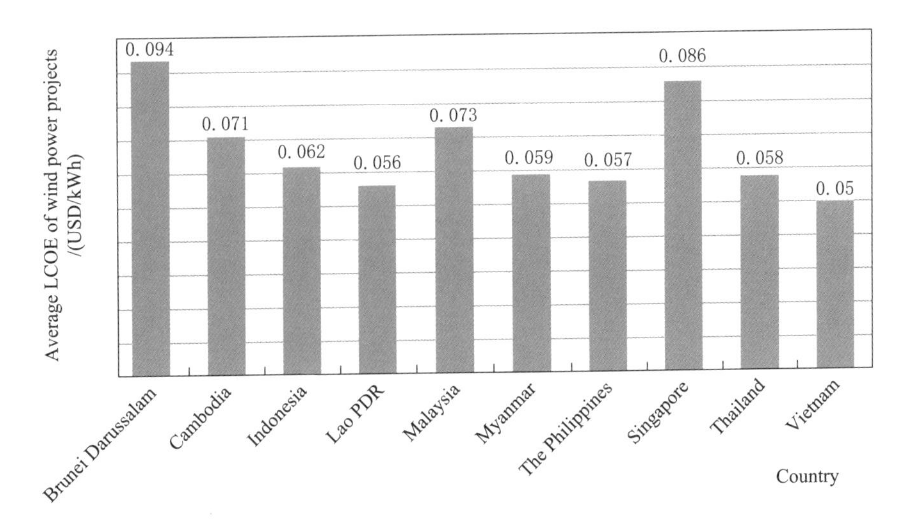

Figure 3. 4. 2 Comparison of average LCOE of wind power projects in AMS

The LCOE for wind power development in the AMS is between 0. 05 USD/kWh and 0. 094 USD/kWh, and the regional average LCOE is 0. 056 USD/kWh. In terms of spatial distribution, the regions with low LCOE are mainly distributed in the central and eastern Indo-China Peninsula, and the central and northern the Philippines. Vietnam, Thailand, Lao PDR, Myanmar, and the Philippines, where the technically exploitable capacity of wind energy is high, have relatively low development cost. Except Vietnam, the average LCOE of the other countries is higher than the global average of onshore wind power LCOE of 0. 053 USD/kWh in 2019. Limited by resource conditions, according to the LCOE calculation in Figure 3. 4. 3, the LCOE in Singapore and Brunei Darussalam are 0. 086 USD/kWh and 0. 094 USD/kWh, respectively. The average wind power development cost of AMS is higher than the world average. An independent wind energy industry supply chain has not yet formed in ASEAN, hence wind turbines and related equipment and materials need to be transported by land or sea, which increases transportation cost.

Box 3. 3 Calculation parameters and methods of wind power LCOE for ASEAN

The calculation formula in Box 3. 2 is used to calculate the LCOE of technologically developable areas in AMS. The wind power generation of each grid is calculated according to the utilization hours of the grid estimated by the wind energy resources and the installation capacity in the grid. The other parameters are shown in Table 3. 4. 1.

Table 3.4.1 Main parameters for calculating LCOE of wind power in AMS

Country	Loan interest rate	Loan ratio	Tax rate	Construction cost/(USD/kW)	O&M cost /[USD/(kW · a)]
Brunei Darussalam	5.5%	70%	18.5%	1,750	32
Cambodia	5.5%	70%	20%	1,700	32
Indonesia	5.5%	70%	25%	1,650	23
Lao PDR	5.5%	70%	25%	1,700	35
Malaysia	5.5%	70%	24%	1,650	35
Myanmar	5.5%	70%	25%	1,750	35
The Philippines	5.5%	70%	20%	1,650	35
Singapore	5.5%	70%	17%	1,600	30
Thailand	5.5%	70%	20%	1,700	23
Vietnam	5.5%	70%	20%	1,500	31

Notes 1. Since current wind power projects usually adopt the loans from international financial institutions with lower interest rates, the loan interest rate and loan ratio are uniformly taken based on the experience of existing projects.

2. The tax rate refers to the relevant research results of NREL and ACE.

3. The construction cost covers the expenses for equipment, machinery, civil works, installation and commissioning, financing and interest in the period of construction, and facilities for connection with network. In Vietnam and Thailand, where the data of completed projects are available, the construction cost is estimated with reference to the actual cost level. In other countries, the construction cost is estimated by comparing the conditions of infrastructure and transportation with that in Vietnam and Thailand.

4. Operation and Maintenance (O&M) cost only cover the expense related to the operation and maintenance of equipment, including equipment maintenance and repair. In the period of operation, the expense of land lease, labour cost, insurance, and etc, are uniformly taken at 2.0% of the initial investment.

5. In this research, the construction cost and O&M cost of wind power plants in a specific country are taken as a single value. However, there are individual differences in the construction cost and O&M cost for specific projects. The estimated LCOE in this report do not reflect this influencing factor.

6. In this research, the estimated LCOE only considers the investment of the wind power plant and does not include any external conditions, such as the cost of transmission lines and national wind power incentive policies, such as, tax deductions and incentives for AMS.

3.4.2 Scenario analysis and future trend study

To analyse the feasible path for the reduction of wind power costs in AMS, the factors that affect LCOE are divided into four categories.

The first is natural factors, which mainly refers to wind energy resource conditions. Projects with better wind energy resources have higher annual utilization hours and larger power generation with the same installed capacity, making projects'cost per kilowatt-hour lower. Therefore, the areas with better wind energy resources are usually preferred for development. For a specific area, the wind resource condition is certain, and the space for reducing development costs from the perspective of wind energy resources is limited.

The second is the economic factors of a country where the project is located, such as the tax rate, inflation rate, and labour cost. These factors will have influence on the discount rate and project Operation and Maintenance (O&M) costs to affect the calculation results of LCOE; but these factors are usually determined by the national economy, and it

is difficult to reduce a project's development cost through some measures taken by a project developer. AMS could formulate supporting policies offering tax reduction and other attractive incentive programs as efforts to promote the development of wind power. This practise has been successfully applied in Vietnam to further reduce the LCOE of wind power generation.

The third is the project construction factors. As one of the main components of construction cost, the price of WTGS still has room for a certain reduction. The project contractor can also reduce the construction cost through optimized management at the construction stage. In addition, with the rapid economic development and the continuous improvement of infrastructure such as ports and transportation, the logistics cost will be further reduced. The readiness of local supply chains could also affect the reduction of LCOE as it will significantly cut cost for transportation of some of the essential equipment in wind power generation.

The fourth is financing factors. The loan interest rate will directly affect the internal rate of return on the investor's capital. LCOE can be lowered effectively by reducing project financing cost through diversified financing channels to make full use of low-interest loans from international financial institutions and grants from international organization. Supporting policies to ease the access of project financing may also act as a catalyst that could further help lower the cost. Therefore, project construction costs and financing costs play an important role in the reducing of LCOE and also have certain room for reduction, so they can be considered as the main direction of wind power project LCOE reduction efforts.

In order to understand the impact of construction costs and loan interest rates on the LCOE of wind power projects, LCOE changes in a combination of different scenarios, where construction costs and loan interest rates have fallen, are analysed, taking Vietnam as an example (as shown in Table 3. 4. 2).

Table 3. 4. 2 Impact of construction costs and loan interest rates on the project LCOE (USD/kWh)

Loan interest rate	Construction cost			
	0	−10%	−20%	−30%
0	0. 050	0. 046	0. 042	0. 038
−10%	0. 048	0. 044	0. 040	0. 036
−20%	0. 046	0. 042	0. 039	0. 035
−30%	0. 044	0. 041	0. 037	0. 034

Based on the average LCOE of current wind power projects in Vietnam, when construction costs remain unchanged, loan interest rates will reduce by 30% and LCOE will decrease by 0. 006 USD/kWh; whereas when the loan interest rate remains the same, construction costs will drop by 30% and LCOE will decrease by 0.012 USD/kWh. Therefore, the LCOE of wind power projects is more sensitive to the change of the construction cost. If the construction cost and the loan interest rate fall by 30% at the same time, the LCOE of wind power projects will be 0.034 USD/kWh, which is 32% lower than the reference scenario and lower than the global average of LCOE 0.041 USD/kWh for wind power projects in the second half of 2020.

In future, with the progress of carbon neutrality, the importance of large-scale development of wind power will become increasingly prominent, and the improvement of policy incentives for the wind power industry in AMS will expand the space for development of wind power. Meanwhile, the constant progress of wind power technology will continue to promote the wind power industry to reduce costs and increase efficiency. Especially considering the construction cost and financing cost, the development costs of wind power projects in AMS still have a certain room for reduction.

Chapter 4

Analysis on LWSP Development Landscape in AMS

4.1 Analysis on External Conditions for LWSP Development

4.1.1 Current situation of power system

4.1.1.1 Power supply and generation

AMS are rich in fossil energy resources, on which the power mix has relied for a long time. In recent years, to fulfil the Paris Agreement and achieve the Sustainable Development Goal (SDG), AMS see renewable energy development as one of their development priorities. As of the end of 2019, the total installed capacity of ASEAN was 251.5 GW, among which solar power of fossil energy was 181.0 GW, accounting for 72.0% of the total; that of hydropower was 50 GW, accounting for 19.9% of the total; and that of other renewable energies, such as wind power, solar power, biomass power, and geothermal power was 20.5 GW, accounting for 8.2% of the total. The installed capacity of wind power was 2.3 GW, accounting for only 0.9% in the whole structure of power supply sources. It is worth noting that the installed capacity of natural gas fired power accounts for 36.1% of the installed power capacity in AMS, which can flexibly adapt to the changes in daytime and seasonal power demand (Figure 4.1.1). High proportion of adjustable natural gas fired power installed capacity provides favourable conditions for development of variable renewable energy, such as wind power and solar power, in the future in AMS.

In 2019, the total energy production of ASEAN was 1,111 TWh, in which the output of fossil energy accounted for 76.6%, including that of gas-fired power plants, coal-fired power plants and oil-fired power plants, 34.5%, 40.2%, and 1.9%, respectively. In addition, the output by renewable energy was 243.7 TWh, accounting for 21.9%, including the hydropower output of 15.6%, and the outputs by other renewable energies, such as wind power, solar power, geothermal power, and biomass power, accounting for 6.3%. The wind power output only accounted for 0.4% of the total energy production (Figure 4.1.2).

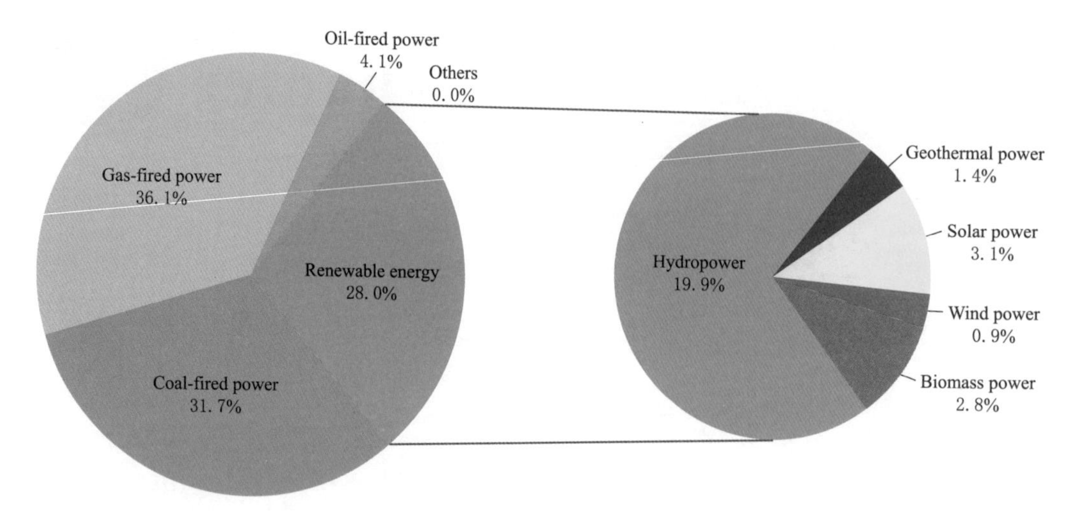

Figure 4. 1. 1 Structure of power supply sources of ASEAN in 2019

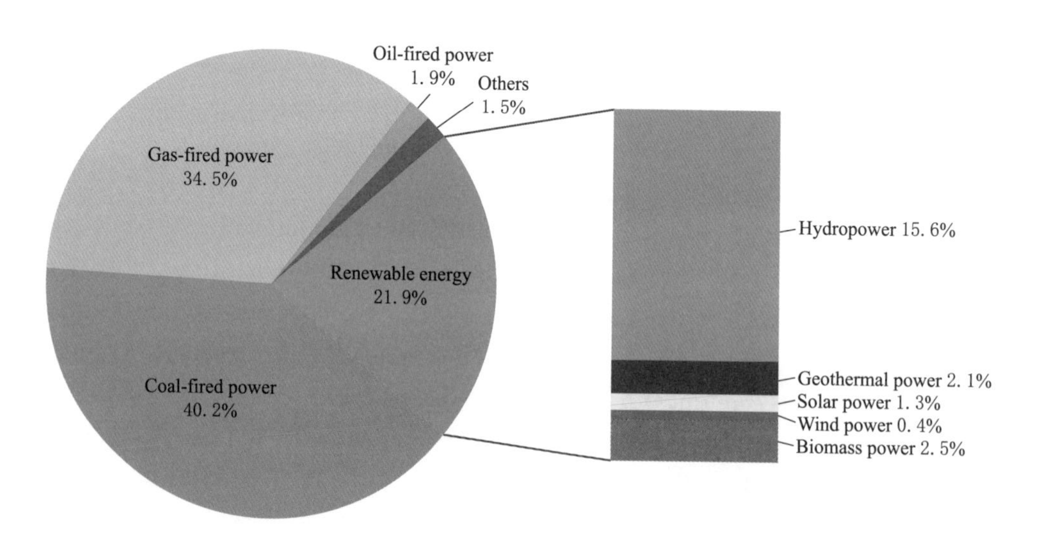

Figure 4. 1. 2 Power generation proportion of various power supply sources of ASEAN in 2019

The installed capacity situation of AMS in 2019 is shown in Figure 4. 1. 3. In Lao PDR, Cambodia, Myanmar, and Vietnam, the proportion of renewable energy is close to or more than 50%, while in the Philippines and Thailand, the installed capacity proportion of other renewable energies (except for hydropower) is relatively high, reaching about 14%, and the wind power capacity accounts for more than 2%.

4. 1. 1. 2 Power grid

Due to geographical location, historical and economic development, etc., the development of power grids is different in AMS. Singapore, Thailand, and Brunei Darussalam

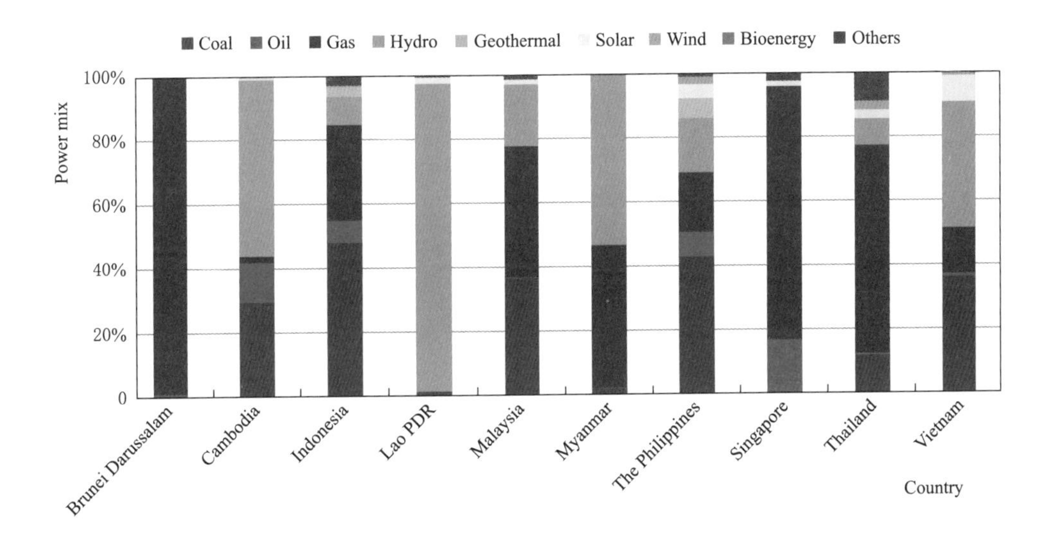

Figure 4. 1. 3　Power installed capacity of AMS in 2019

have built relatively perfect power grids, with the rate of access to electricity up to (or approaching) 100%. However, Myanmar and Cambodia are lagging in power grid construction. As of 2019, about 50% of the population in Myanmar lacked power supply, and the power connection rate in Cambodia was only about 75%. In addition, the Philippines and Indonesia have not yet established a unified power network.

The high voltage transmission network in Thailand consists of 500 kV, 230 kV, 132 kV, 115 kV, and 69 kV lines, which is uniformly operated by Electricity Generating Authority of Thailand (EGAT). EGAT dispatches the power network through the National Control Center (NCC) in Bangkok and five regional control centers (RCC) in the north, central part, south, east, and northeast, to meet the power demand in the whole country.

Lao PDR's power grid system is divided into four parts in regions, namely, northern power grid, central power grid Ⅰ, central power grid Ⅱ, and southern power grid, including three types: high voltage (500 kV, 230 kV, 115 kV), medium voltage (22 – 35 kV) and low voltage (0. 4 – 12. 7 kV). The system mainly adopts 115 kV lines, and the 500 kV transmission line is less than 100 km long in total, mainly for cross-border power trade between Lao PDR and Thailand. Lao PDR's power grid is weak in risk resistance. According to the statistics, the power grid has collapsed for 16 times since 2020. In addition, the recovery measures are single, and a long time is needed to recover, generally more than 3 hours.

Since there are a lot of islands in the Philippines, its power grid is mainly composed of three parts, located in Luzon, Visayas, and Mindanao. Due to geographical constraints,

the power grids are radial in most areas, and at its highest voltage level, the 230 kV is mostly composed in mesh-loop and only in Manila on Luzon Island. The main islands are connected by Alternating Current/Direct Current (AC/DC), and for the line between islands, it is mainly connected by submarine cable connecting lines. Since most down-line within the island is mainly radial, the backbone grid is relatively weak. This situation exposes high risk of frequent power outages. Deploying LWSP will increase number of inertia-less power source, in which careful study is necessary to be conducted in this context.

Vietnam has established a national power grid supported by a 500 kV backbone grid and connecting the northern, central, and southern power systems. The power systems are operated and dispatched by different levels, that is the national load dispatching centre, regional load dispatching centre, and local load dispatching centre, which are responsible for 500 kV ultra-high voltage, 220 - 110 kV high voltage, and 35 kV (and below) medium and low voltage power systems, respectively. As of 2017, Vietnam power grid is equipped with 27 substations with 500 kV ultra-high voltage and a capacity of 29,400 MW and 500 kV transmission line of 7,445 km, and 220 kV substations with a capacity of 48,053 MW and 220 kV transmission line of 17,010 km. In Vietnam, the structure of southern power grid is better than that of the northern part due to its greater transmission capacity. In recent years, the PV industry in Vietnam has developed rapidly, and it is difficult for the existing transmission and distribution lines to tap the new solar power generation. Under such situation, Vietnam has formulated and implemented relevant supporting policies. For instance, in mid - 2020, the Ministry of Industry and Trade, Vietnam has proposed a new Public-Private Partnership (PPP) Bill which allows private companies to invest in transmission lines and substations connecting power plants with the national grid.

Malaysia is divided into East Malaysia, the Sabah and Sarawak region, and West Malaysia, the Peninsular region, with a distance of more than 600 km. Therefore, the power grid is also divided into three independent power systems. The power system in West Malaysia is managed by Tenaga Nasional Berhad (TNB), while the power system in East Malaysia is managed by Sabah Electricity Sdn Bhd (SESB) and Sarawak Energy Berhad (SEB). Under the development of the APG, there will be connecting lines between East and West Malaysia. Tapping such potential will allow mobilisation of renewable energy, which is abundant in the east region, to the high growing demand in the west region. A careful feasibility study would unlock more potential to deploy LWSP, particularly in the west region.

In Indonesia, although the single buyer system is adopted, a unified power system has not yet been built. The interconnectivity of the power grid in Indonesia is low due to the large number of islands. The national power system consists of more than 600 independent sub-grids and 8 main regional power grids, with a distributed structure, and mainly covers the densely populated areas such as Java, Bali, Sumatra, and Kalimantan Island. For a long time, only Java, Madura, and Bali have achieved interconnection of power grids, with the name Jamali Grid. As for Sumatra region, the interconnection within the island has already been formed but not yet interconnected to the Jamali Grid. Moreover, most of the existing Sumatran power plants are exclusively connected to a particular group of grids. Thus, most of the existing power plants can only supply power to surrounding areas.

In the region, ASEAN's energy resources are distributed unevenly, and the current existing infrastructures within ASEAN region do not satisfy the requirement for regional interconnection. Lao PDR and Myanmar have abundant hydropower resources and small power demand, while Thailand and Vietnam have large power demand and need to import electric power. It is the consensus of AMS to strengthen regional power interconnection and coordinate the allocation of regional energy and power resources. For this reason, ASEAN put forward the ASEAN Power Grid (APG) Program in 1997, which aimed at integrating ASEAN power system, formulating APG rolling development planning, and promoting it as a common core project for economic development of AMS. The *ASEAN Plan of Action for Energy Cooperation* (*APAEC*) 2016 – 2025 *Phase* Ⅰ: 2016-2020 formulated in 2015 clarified that 16 bilateral power interconnection projects with a total of 45 transmission channels should be promoted in this phase. At present, nine cross-border power transmission projects have been completed, with 20 power interconnection channels and an interconnection capacity of about 7,720 MW. After the completion of all projects, the power transmission capacity among AMS can be as high as 26,000 – 30,000 MW. In 2020, ASEAN continuously launched the *APAEC* 2016 – 2025 *Phase* Ⅱ: 2021 – 2025, which pointed out that AMS would continue the devotion to accelerate the development of APG and enhancing the deployment of renewable energy through investment and financing. To this end, the electric power companies and competent authorities of AMS have adopted the strategy for accelerating the progress of APG projects and expand multilateral power trade, and conduct a research to determine the feasibility of at least four APG projects in the next step.

As the main direction of APG, multilateral power trade is conducive to improving the flexibility of power systems in AMS, to integrate the renewable energy in a higher proportion. In addition, the multilateral power trade can also improve the system security and economic efficiency, and help achieve the target of renewable energy development in line with the decarbonization agenda. For example, Vietnam has established cross-border power grid interconnection with its neighbouring countries, China and Lao PDR. At present, Vietnam imports more than 450 MW of electric power from China and more than 500 MW from Lao PDR. Vietnam plans to increase the electric power imported from China to 3,000 MW in the coming decades, so it is expected that the power transmission capacity will increase in the future.

4.1.2 Estimation of power demand

By analysing the social and economic development, the medium and long-term power demand is estimated for ASEAN region.

4.1.2.1 Social and economic development status

The statistical data of Gross Domestic Product (GDP), population, and electricity consumption for 10 AMS in the period of 2011 to 2020 are shown in Table 4.1.1, from which it can be seen that the overall growth rate of GDP was at a high level in the past 10 years and became-4.4% in 2020 due to global epidemic, and the population growth rate declined from 1.95% in 2011 to about 1.1% in 2020. Besides, the average annual growth rate of electricity consumption was about 5.7% and the average annual growth rate of electricity consumption per capita was about 4.4%. In 2020, affected by the global epidemic, the electricity consumption increased by only 1.81% compared with 2019, and the electricity consumption per capita increased by only 0.71%, far below the historical average. The electricity elasticity coefficient varies in a wide range, and the average value is about 0.84 in recent 10 years.

Table 4.1.1 Statistic of GDP, population, and electricity consumption of AMS
(Source: ASEAN Energy Database System, AEDS)

Year	2011	2012	2013	2014	2015	2016	2017	2018	2019	2020
GDP/(trillion USD)	2.26	2.40	2.51	2.55	2.47	2.60	2.81	3.00	3.17	3.04
GDP growth rate/%	16.53	6.34	4.59	1.28	−3.03	5.17	8.03	6.82	5.67	−4.0
Population/(100 million persons)	5.97	6.05	6.12	6.21	6.28	6.35	6.42	6.49	6.56	6.63
Population growth rate/%	1.95	1.32	1.27	1.32	1.21	1.15	1.10	1.06	1.07	1.10
Electricity consumption /TWh	615	667	704	744	792	852	888	940	992	1,010
Growth rate of electricity consumption/%	4.05	8.49	5.60	5.57	6.43	7.68	4.21	5.81	5.55	1.81
Electricity consumption per capita/kWh	1,030	1,103	1,150	1,199	1,260	1,342	1,383	1,448	1,512	1,523
Growth rate of electricity consumption per capita /%	2.06	7.08	4.27	4.20	5.16	6.46	3.08	4.71	4.43	0.71
Electricity elasticity coefficient	0.24	1.34	1.22	4.36	−2.12	1.49	0.53	0.85	0.98	−0.45

4.1.2.2 Estimation of power demand by elasticity coefficient method

As AMS have developed rapidly in society and economy in recent years, this research, referring to the research result of ACE, assumes that GDP growth rates will be 5.15%, 5.10%, 4.6%, and 4.4%, respectively, in 2025, 2030, 2040, and 2060.

The elasticity coefficient for electricity consumption is closely related to the growth rate of GDP. The average value of ASEAN in the past 10 years is around 0.84, but the elasticity coefficient is negative because of the negative growth of GDP in 2015 and 2020. After excluding the two abnormal years, the average value is around 1.38. However, considering the elasticity coefficient abnormally large because of a small growth of GDP in 2014, and combined with the general law of elasticity coefficient, the elasticity coefficient of ASEAN can be taken as about one. Considering the strong driving force of ASEAN's future economic growth, with the acceleration of industrialization, the enhancement of power supply rate, and the improvement of people's living standards, the elasticity coefficient will increase steadily in the next five years. With economic growth at a certain level, the elasticity coefficient of electricity consumption will gradually decline after 2026. The elasticity coefficient of electricity consumption is taken as 1.1 from 2020 to 2025, 1.0 from 2026 to 2030, 0.85 from 2031 to 2040, and 0.8 from 2041 to 2060. Table 4.1.2 presents the power demand of AMS estimated by the elasticity coefficient method.

Table 4.1.2 Estimation of power demand of ASEAN (by elasticity coefficient for electricity consumption)

Item	Unit	2020	2025	2030	2040	2060
Electricity consumption	TWh	1,010	1,330	1,706	2,504	5,001
Growth rate	%	1.81	5.67	5.1	3.91	3.52
Elasticity coefficient		0.35	1.1	1	0.85	0.8
GDP growth rate	%	5.2	5.15	5.1	4.6	4.4

4.1.2.3 Estimation of power demand by electricity consumption per capita

Based on the trend of population growth in AMS over the years and by reference to the research results of ACE, it is assumed that the population growth rate of AMS will be 0.99% in 2025, 0.77% from 2026 to 2030, 0.55% from 2031 to 2040, and 0.5% from 2041 to 2060.

With the continuous social and economic development of AMS, the electricity consumption per capita has increased from 1,030 kWh/person in 2011 to 1,523 kWh/person in 2020, with an average annual growth rate of 4.44%. However, compared with the developed countries in Europe and America, there is still a big gap, and the electricity consumption per capita of ASEAN will increase rapidly in recent decades. It is expected that the growth rate of electricity consumption per capita will be 5%-3.6% from 2025 to 2060. Based on the growth rate, it is expected that the electricity consumption per capita of ASEAN will reach 3,637 kWh/person in 2040, the same as that of China in 2012, and be up to 7,378 kWh/person in 2060, equivalent to that of Japan in 2016. The calculation results by electricity consumption per capita are shown in Table 4.1.3.

Table 4.1.3 Estimation of power demand of ASEAN (by electricity elasticity per capita)

Item	Unit	2020	2025	2030	2040	2060
Electricity consumption	TWh	1,010	1,354	1,779	2,781	6,234
Population	100 million	6.63	6.97	7.24	7.65	8.45
Electricity consumption per capita	kWh/person	1,523	1,944	2,457	3,637	7,378

4.1.2.4 Selection of estimated results

As a whole, the estimated result by the electricity consumption per capita method is larger than that by the elasticity coefficient method. Considering that there is still much room for growth in ASEAN's power demand in the future, and that the historical data adopted by the electricity elasticity coefficient method is not very regular, it is recommended to adopt the estimated result by the electricity consumption per capita method in this research. In such case, the power demand of ASEAN in 2025, 2030, 2040, and 2060 will be 1,354 TWh, 1,779 TWh, 2,781 TWh, and 6,234 TWh, respectively, and the power demand estimation results from 2025 to 2040 are close to the general development scenarios in the research results of ACE.

4.1.3 Power development plan potential

Based on the installed capacity of ASEAN power system in 2019, the balance of energy and the balance of power are analysed, with the results as shown in Tables 4.1.4 and 4.1.5. Based on the existing installed capacity of ASEAN, the ASEAN's energy shortage will be 405 TWh, 881 TWh, 2,004 TWh, and 5,871 TWh, respectively, and the power shortage will be 79 GW, 174 GW, 387 GW, and 1,143 GW, respectively, in 2025, 2030, 2040, and 2060. Considering that the existing power mix of ASEAN in 2019 remains unchanged, it is needed to increase the installed capacity by at least 99 GW, 217 GW, 481 GW, and 1,424 GW in 2025, 2030, 2040, and 2060, respectively, so as to ensure the balance between supply and demand of power and energy.

Table 4.1.4 Electric energy balance analysis of ASEAN power system unit: TWh

No.	Item	2019	2025	2030	2040	2060
1	Power demand	1,111	1,517	1,992	3,115	6,983
1.1	Electricity consumption	992	1,354	1,779	2,781	6,234
1.2	Loss	119	162	213	334	748
2	Generating capacity	1,111	1,111	1,111	1,111	1,111
2.1	Gas-fired power	383.6	384	384	384	384

Continued

No.	Item	2019	2025	2030	2040	2060
2.2	Coal-fired power	446.7	447	447	447	447
2.3	Oil-fired power	20.6	21	21	21	21
2.4	Hydropower	173.4	173	173	173	173
2.5	Wind power	3.9	4	4	4	4
2.6	Solar power	14.5	14	14	14	14
2.7	Biomass power	28.1	28	28	28	28
2.8	Geothermal power	23.8	24	24	24	24
2.9	Others	16.8	17	17	17	17
3	Electric energy surplus (+) / loss (−)	0	−405	−881	−2,004	−5,871

Table 4.1.5 Electric power balance analysis of ASEAN power system

Item	Unit	2019	2025	2030	2040	2060
Ⅰ. Electricity consumption	TWh	992	1,354	1,779	2,781	6,234
Ⅱ. Utilization hours at maximum load	h	5,400	5,300	5,200	5,200	5,100
Ⅲ. Capacity required by system	GW	202	281	376	588	1,345
Maximum load	GW	184	255	342	535	1,222
Reserve capacity	GW	18	26	34	53	122
Ⅳ. Total installed capacity	GW	251	251	251	251	251
Gas-fired power	GW	90.9	91	91	91	91
Coal-fired power	GW	79.8	80	80	80	80
Oil-fired power	GW	10.3	10	10	10	10
Hydropower	GW	50.0	50	50	50	50
Wind power	GW	2.2	2	2	2	2
Solar power	GW	7.7	8	8	8	8
Biomass power	GW	7.0	7	7	7	7
Geothermal power	GW	3.5	3	3	3	3
Others	GW	0.0	0	0	0	0
Ⅴ. Available installed capacity	GW	201.8	201.8	201.8	201.8	201.8
Gas-fired power	GW	81.8	81.8	81.8	81.8	81.8
Coal-fired power	GW	71.8	71.8	71.8	71.8	71.8
Oil-fired power	GW	9.3	9.3	9.3	9.3	9.3
Hydropower	GW	30.0	30.0	30.0	30.0	30.0
Wind power	GW	0.4	0.4	0.4	0.4	0.4
Solar power	GW	1.5	1.5	1.5	1.5	1.5
Biomass power	GW	5.6	5.6	5.6	5.6	5.6
Geothermal power	GW	1.4	1.4	1.4	1.4	1.4
Others	GW	0.0	0.0	0.0	0.0	0.0
Ⅵ. Electric power surplus/ loss	GW	0	−79	−174	−387	−1,143

Under the background of global energy crisis and environmental crisis, it has become an inevitable trend of power development in the future to replace the traditional fossil energy by clean renewable energy, and the proportion of output by renewable energy to the total outputs will gradually increase. As the utilization hour of power generated by renewable energy is generally lower than that by traditional fossil energy, the installed power capacity of renewable energy in ASEAN will be further scaled up, therefore, there is a huge room for development of renewable energy in AMS in the future.

4.1.4 Market environment and supporting policies

To support the development of renewable energy, AMS have promulgated relevant policies on regulation, financial incentives and public financing, whereas the specific situation is summarized as shown in Table 4.1.6. Up to now, the Philippines, Thailand, Vietnam, and Indonesia have all put forward the goals of renewable energy development, which involve the development and deployment of wind power industry. In addition, Malaysia and Cambodia have implemented the policies for FiT and tax preference to encourage the development of renewable energy.

In regard to the goals of renewable energy development in AMS, by the end of 2019, renewable energy accounted for 12.1% of the primary energy supply and 28% of the installed capacity; according to the *APAEC 2016 - 2025 Phase* Ⅱ: 2021 - 2025, ASEAN has updated the renewable energy development goals. By 2025, the renewable energy will account for 23% of the TPES and 35% of the installed power generation capacity. Meanwhile, AMS have also formulated their own renewable energy development goals and plans, as shown in Table 4.1.7.

In terms of climate change goals, AMS have all made Nationally Determined Contribution (NDC) under the framework of the Paris Agreement and pledged to implement the Sustainable Development Goal (SDG) and support clean energy development. Policies in the two aspects are summarized as shown in Table 4.1.8.

Overall, AMS have formulated the related goals and policies for renewable energy development, but the support for wind power industry varies to a great extent among different countries. Thailand, Vietnam, the Philippines, and Indonesia have made the clear planning for wind power development and promulgated the incentive policies such as FiT and tax preference. The wind power industry has been initiated there, in which the wind power development of Thailand, the Philippines, and Vietnam are satisfactory. Although there is a certain potential for wind power development in Myanmar, Lao PDR, and Cambodia, vigorous industry supportive policies are lacking, and the wind power industry development is in urgent need of breakthroughs. Restricted by wind energy resources, Brunei Darussalam, Singapore, and Malaysia do not make wind power development a priority.

Table 4. 1. 6 **Renewable energy policies of AMS**

Policy type		The Philippines	Vietnam	Indonesia	Malaysia	Thailand	Singapore	Myanmar	Lao PDR	Cambodia	Brunei Darussalam
Regulatory policy	Renewable energy target involved in INDC or NDC	•	•	•	•	•	•	•	•	•	•
	Renewable energy target	•	•	•	•	•	•	•	•		•
	Feed-in Tariff/auction/ premium mechanism	•	•	•	•	•				•	
	Net metering/ billing/direct consumption	•	•	•	•		•				
	Biofuel mixture consumption obligation/task/target	•	•	•	•	•					
	Power facilities quota/renewable energy quota system (RPS)	•	•	•	•						
	Renewable energy certificate (REC) Transaction	•	•	•	•	•					
	Renewable energy heat supply quota/target		•								
Financial incentive and public financing	Tax preference	•	•	•	•	•		•	•	•	
	Public investment/loan/grant/ subsidy	•	•	•	•	•	•		•		
	Sales tax, value-added tax, or other tax relief	•	•	•	•	•		•			
	Bid	•		•	•		•				
	Tax credit for investment or power generation	•	•	•							
	Payment for power generation	•				•					

Table 4.1.7 Targets of renewable energy development in AMS

Country	Target year	Target
ASEAN	2025	Renewable energy accounts for 23% of TPES, and 35% of the installed capacity
Brunei Darussalam	2035	Renewable energy accounts for 10% in the power generation mix, i.e., 954,000 MWh
Cambodia	2030	Hydropower accounts for 55% of the generation portfolio, and other renewable energy* accounts for 10% * *Other renewable energy: biomass accounts for* 6.5% *and solar energy takes up* 3.5%
	2035	Rooftop solar PV power meets 3% of the residential power demand
Indonesia	2025	Renewable energy accounts for 23% of TPES (92 million tons of petroleum equivalent) The installed capacity of renewable energy is up to 45 GW 20% bioethanol mixing target Maintain biodiesel mixing ratio at 30%
	2050	Renewable energy accounts for 31% of TPES (310 million tons of petroleum equivalent) The installed capacity of renewable energy reaches 168 GW 50% bioethanol mixing target Maintain biodiesel mixing ratio at 30%
Lao PDR	2025	Renewable energy accounts for 30% of TFEC (excluding large-scale hydropower), including 20% of output by renewable energy (excluding large-scale hydropower), and 10% by biofuel (mixing ratio at 5%-10%)
Malaysia	2025	Renewable energy accounts for 31% of the installed capacity portfolio
	2035	The deployment of battery energy storage system reaches 500 MW
Myanmar	2025	Renewable energy accounts for 20% of the installed capacity portfolio (excluding large-scale hydropower exceeding 100 MW)
	2030	Renewable energy accounts for 9% of the total installed capacity (excluding small hydropower) Hydropower accounts for 38% of the total installed capacity Renewable energy accounts for 12% of the national generation portfolio (excluding large-scale hydropower)
The Philippines	2019	Biodiesel mixing at 2%, bioethanol mixing at 10%
	2030	The installed capacity of renewable energy triples from 5.4 GW in 2010 to 15.3 GW in 2030
	2040	The installed capacity of renewable energy reaches 20 GW in 2040
Singapore	2020	Achieve the solar energy target of 350 MWp
	2025	Achieve the solar energy target of 1.5 GWp
	2030	Achieve the solar energy target of at least 2 GWp
	After 2025	Achieve the energy storage deployment target of 200 MW

Continued

Country	Target year	Target
Thailand	2037	Renewable energy accounts for 30.18% of TFEC, of which the output by renewable energy accounts for 5.75% of the total outputs, renewable energy takes up 21.2% of heat consumption, and biofuels account for 3.22% of TFEC Power generation by renewable energy by 2037 • Solar energy: 12,139 MW • Floating solar energy: 2,725 MW • Biomass: 5,790 MW • Wind power: 2,989 MW • Large-scale hydropower: 2,920 MW • Small-scale hydropower: 308 MW • Biogas (wastewater/livestock excrement/energy crops): 1,565 MW • Municipal solid waste: 900 MW • Industrial waste: 75 MW
Vietnam	2030	Renewable energy accounts for 15%-20% of TPES (in which wind power takes up to 16,000 MW)
	2045	Renewable energy accounts for 25%-30% of TPES (in which wind power takes up to 39,610 MW)

Table 4.1.8 Summary of NDC and SDG policies of AMS

Country	NDC goal		SDG 7 progress	
	Year	Goal	Electrification rate	Proportion of renewable energy
Brunei Darussalam	2035	Total energy consumption is reduced by 63%	100%	Renewable energy accounts for 0.05% of the total outputs
Cambodia	2030	Greenhouse gas emissions are reduced by 27%	93%	Renewable energy accounts for 62% of the total energy consumption (mostly hydropower)
Indonesia	2030	Greenhouse gas emissions are reduced by 29%	99%	Renewable energy accounts for 8.8% of the total outputs (mostly hydropower and geothermal power)
Lao PDR	—	The amount of carbon dioxide generated in transportation is reduced by 191,000 tons of carbon dioxide equivalent per year	100%	Renewable energy accounts for 83% of the total outputs (mostly hydropower)
Malaysia	2030	Greenhouse gas emissions per GDP are reduced by 35%	100%	Renewable energy accounts for 14% of total outputs
Myanmar	2030	Electric power is saved by 20%	68%	Renewable energy accounts for 61% of the total energy structure
The Philippines	2030	Greenhouse gas emissions are reduced by 75%	96%	Renewable energy accounts for 68.1% of the total energy structure (mostly geothermal power)

Continued

Country	NDC goal		SDG 7 progress	
	Year	Goal	Electrification rate	Proportion of renewable energy
Singapore	2030	Peak emissions at 65 $MTCO_2e$ around 2030	100%	0.6% of renewable energy share in total energy mix
Thailand	2030	Greenhouse gas emissions are reduced by 20%	100%	Renewable energy accounts for 15% of the total energy consumption and 10% of the total outputs
Vietnam	2030	Greenhouse gas emissions are reduced by 9%	100%	Renewable energy accounts for 35% of the total outputs (mostly hydropower)

Brunei Darussalam: As Brunei Darussalam is rich in oil and gas resources, hence they are still prioritizing their policy on oil and gas. At present, there is no clear scheme or policy for FiT of renewable energy. Brunei Darussalam is endowed with abundant solar energy resources, so Brunei Darussalam may introduce FiT and net metering policies in the future to encourage the investment in solar PV power.

Cambodia: Considerable incentive policies for renewable energy development, such as FiT and tax preference, have been promulgated and implemented. In addition, some incentive measures have also been taken, for example, providing investors with land, infrastructure, financial mechanism, etc. Although wind energy resources are of certain development potential, the commercial access procedures of power projects in Cambodia are complicated, which results in a few investments in wind power industry. In such case, the wind power development is still at its initial stage. At present, the Cambodian Government is negotiating with The Blue Circle to develop the first wind farm.

Indonesia: According to the National Energy Plan (RUEN), Indonesia has set a target to reach 1.8 GW of wind power generation by 2025, and 28 GW by 2050. Incentive policies and measures for renewable energy have been promulgated, and specific planning and development roadmap have also been made for wind power industry, but the installed capacity of wind power is far less than that of fossil energy power. In terms of the specific policies, the Indonesian Government has implemented a FiT of 17 cents/kWh for wind and tidal power. According to the capacity quota plan published by the government, wind power plants should be purchased by the developers through auction.

Lao PDR: Up to now, the grid-connected renewable energy systems have not implemented the FiT. However, the state provides tax preference and financial support for domestic and foreign investors in renewable energy projects (including wind power investment). Thailand Renewable Energy Group (BCPG) and IEA signed an agreement to establish the ASEAN's largest wind farm in Lao PDR.

Malaysia: Comprehensive policy incentives for renewable energy, regulatory support, pricing mechanism, and interconnection procedures have always been driving the development of renewable energy in Malaysia. Malaysia has formulated FiT policies for renewable energies such as biomass, biogas, solar energy, and small hydropower. Due to Malaysia's geographical location, the onshore wind power development is restricted due to the low technical potential and economic feasibility. Therefore, wind power development is not made a priority by the Malaysian Government, and there is also no FiT policy on wind power.

Myanmar: Investors in Myanmar's renewable energy industry can enjoy various preferential policies such as tariff reduction, but there is no specific FiT and other incentive policies on wind power. Considering that Myanmar is of certain potential to develop wind energy resources, it is possible for Myanmar Government to encourage wind power development in the future.

The Philippines: According to the National Renewable Energy Program (NREP), the Philippines has set the goal of achieving the total installed capacity of wind power of 2,378 MW by 2030. The Philippines encourages private capital to invest in new energy and plans to invest 9-10 billion in new energy field in the next 10 years, so as to achieve the goal of doubling the output by new energy. In addition, the Philippines strives to become the largest wind power producing country in Southeast Asia.

The Philippines has introduced several incentive mechanisms in the field of renewable energy, including exemption from income tax for seven years, duty-free import of renewable energy technology and equipment, and FiT scheme applicable to four renewable energies including wind energy. The Philippines implements the wind power FiT system of 15 cents/kWh. Financial institutions of the government provide preferential full-cycle financial services for the development, utilization, and commercialization of renewable energy projects according to suggestions and assessments. In addition to the above incentive measures, the Philippines has also formulated the Renewable Portfolio Standard (RPS), which stipulates that before 2030, power suppliers must acquire the agreed energy supply ratio from eligible renewable energies, with an annual increase of at least 1%.

Singapore: The government has streamlined policies and regulations to facilitate the adoption of renewable energy and make it easier for consumers to be paid for selling excess solar electricity into the grid. The government does not provide a FiT scheme but instead supports renewable energy in the form of research and development solutions that will enable better management of renewable energy intermittency challenges.

Thailand: Since 2004, Thailand has been vigorously supporting the development of renewable energy by government policies and investment incentive measures. Thailand has set the goal of increasing the installed capacity of wind power generation by 3,002 MW by 2036. In addition, the Thai Government has promulgated several incentives and mechanisms, including FiT, auction system, investment subsidy, and board of investment (BOI) incentives (tax exemption period, import tariff reduction, and the right of foreign investors to possess land ownership). Thailand implements the wind power FiT system of 19 cents/kWh. Due to limited resources, Thailand is in urgent need of reducing its dependence on natural gas import, and the development of renewable energy such as wind power has become a priority of the Thai Government.

Vietnam: Vietnam has set the goal of using wind power to supply electric power of 6,000 MW by 2030. In terms of wind power policies, in September 2018, the Vietnamese Government adjusted the FiT of wind power, increasing the FiT of onshore wind power from 7.8 cents/kWh to 8.5 cents/kWh and setting the FiT of offshore wind power to 9.5 cents/kWh. The new FiT has accelerated the development of wind power projects in Vietnam. In addition, wind power enterprises also enjoy the reduction of import tariff, corporate income tax, land tax, and environmental protection fee.

The Vietnamese Government has set a high goal, accompanied by policies for increasing the FiT of wind power and reducing taxes, which has promoted the rapid development of wind power industry. In recent years, the installed capacity of wind power has increased significantly.

Box 4.1 Latest development trend of wind power industry in AMS

A private investor group from Vietnam plans to build 3.4 GW and 3.5 GW offshore wind farms in Thang Long and La Gan. It is expected that the two wind farms will be put into commercial operation before 2030.

The Philippines is working with the World Bank Group to develop an offshore wind power roadmap. At present, the Philippines is conducting a feasibility study on an offshore wind power project with a total capacity of 1,850 MW.

Lao PDR has signed a PPA with Vietnam Electricity (EVN) on a 600 MW Monsoon Wind Power Project in Sekong Province in the south of Lao PDR. In addition, there are seven projects with a total installed capacity of 2,562 MW under feasibility study.

Myanmar plans to build a wind farm in Gunkul to improve the electrification rate. The project is expected to be commenced in 2024, with a total installed capacity of 2,000 MW.

Thailand and Indonesia are expanding onshore wind farms located in Theparak and Sukabumi, respectively.

Considering that there are many sites suitable for wind power development in Cambodia, Cambodia also plans to deploy wind power generation pilot projects. In such case, the installed capacity of wind power may usher in a breakthrough soon.

4.1.5 Opportunities and challenges

4.1.5.1 Opportunities

(1) Addressing climate change and carbon neutrality is pushing ASEAN to develop wind power. Located in one of the worst climate change-affected regions in the world, AMS, have ratified the Paris Agreement and communicated their NDCs to fulfil their responsibilities for combating climate change. Meanwhile, as more and more countries have made clear the time point of carbon neutrality, AMS are also actively exploring and studying the feasible plans for carbon neutrality. AMS urgently need to fully tap the potential of renewable energy including wind power by enhancing the support for renewable energy and accelerating the deployment of renewable energy industry, to reduce the use of fossil fuels, reduce greenhouse gas emissions, help achieve the transition to clean energy and power, and lay the foundation for carbon neutrality.

(2) Advances in technology and globalization of industrial chains can provide opportunities for wind power development in ASEAN. With the continuous improvement of wind power technology, the wind power may be efficiently used in low-wind-speed areas. Due to the globalization of wind power industry chain and especially the signing of Regional Comprehensive Economic Partnership Agreement (RCEP), the import and export tax relief of wind power equipment cut the cost of WTGS, thus the development cost of wind power projects is effectively saved. Technological progress and industry chain globalization have effectively promoted the cost reduction and improved the economic efficiency of wind power development in low-wind-speed areas, thus bringing new opportunities for wind power development in AMS.

(3) Policy planning and promotion of existing projects can bring a good start for wind power development in ASEAN. Up to now, all AMS have set the medium and long-term goals for renewable energy development, among which Thailand, Vietnam, and the Philippines have defined the relevant wind power development goals. Lao PDR, Cambodia, Indonesia, and Myanmar with a certain potential for wind power development have also begun to deploy and promote the development of wind power projects. Under the pressure of energy transition, AMS have been fully aware of the importance of wind power development and begun to promote the development of wind power industry by formulating policy and goal or implementing specific projects based on their own advantages.

4.1.5.2 Challenges

(1) Strengthen the formulation and implementation of planning goals and incentive policies. Some countries have started to promote the development of wind power projects and are expected to achieve breakthroughs in wind power projects in the short-term. However, in the long run, the determination of planning goals and the implementation of sound incentive policies are effective guarantees for attracting investments in wind power development.

(2) Continue to reduce wind power project costs through technological advances and industrial chain development. At present, the construction costs of wind power projects in AMS are slightly higher than the overall international average level. Therefore, it is necessary to further reduce costs and increase the competitiveness of wind power projects against fossil energy projects, by introducing and absorbing advanced technologies related to LWSP development and making full use of location advantages to lay out the wind power industry.

(3) Improve power grids and wind power connection conditions. Due to the large variability, the voltage and frequency of a power grid may be affected in a certain degree when wind power or PV power connects to the power grid. It is necessary to further improve the regional power grids, upgrade the substations and distribution systems, enhance the interconnection of regional power grids, and ensure the newly developed wind power projects are connected to the power grid smoothly. At the same time, researches on the flexibility of power system need to be made to improve its flexibility and prepare for the subsequent expansion of wind power scale by means and measures, such as hybrid energy system of hydropower, wind power and solar power and construction of pumped storage power stations.

4.2 Development Scenario Analysis

To analyse the market space of wind power development in AMS, in this research, the wind power development in AMS in the medium and long-term is expected in two scenarios, i. e., "APAEC Target Scenario" as the development goal and "Leap-forward Scenario". The planning base year is 2019, and the planning level years are 2025, 2030, 2040, and 2060.

(1) APAEC Target Scenario. This scenario is based on the existing renewable energy development goals of AMS, and does not consider the major changes of energy structure and policy environment in the future. According to the development goal set by AMS in the *APAEC* 2016-2025 *Phase* Ⅱ: 2021 – 2025, by 2025, renewable energy will account for 23% of the TPES, and 35% of the total installed capacity. Based on this goal, the penetration rates of wind power generation in AMS in 2025, 2030, 2040, and 2060 are set at 2%, 3%, 4%, and 5%, respectively. By 2060, the penetration rate will be equivalent to the current status of China.

(2) Leap-forward Scenario. In this scenario, the increasing efforts in wind power development is an important measure to tackle climate change and promote the adjustment of energy mix. AMS will formulate more active policies to facilitate the large-scale development of wind power industry, thus laying the foundation for accelerating the realization of carbon neutrality goal. For this scenario, the penetration rates of wind power generation in 2025, 2030, 2040, and 2060 are set at 5%, 6%, 8%, and 10%, respectively. By

2060, the penetration rate will be equivalent to the international medium level at present.

(3) Goal calculation. Combining the power demand of ASEAN in the future, through calculation, the wind power development scale in planning level year under the two development scenarios can be achieved, with the calculation results as shown in Table 4.2.1.

Table 4.2.1 Planning goal of wind power generation in ASEAN

Item		Unit	2025	2030	2040	2060
Power Demand	Electricity consumption demand	TWh	1,354	1,779	2,781	6,234
	Generating capacity demand (taking the loss as 3%-10%)	TWh	1,490	1,930	2,865	6,421
APAEC Target Scenario	Proportion of renewable energy output	%	26	28	30	34
	Proportion of installed capacity of renewable energy	%	35	39	42	47
	Wind power output	TWh	30	58	116	320
	Wind power penetration rate (proportion of output)	%	2	3	4	5
	Installed capacity of wind power	GW	15	27	53	139
	Proportion of installed capacity of wind power	%	4	5	7	8
Leap-forward Scenario	Proportion of renewable energy output	%	27	31	34	39
	Proportion of installed capacity of renewable energy	%	37	43	47	53
	Wind power output	TWh	45	116	228	642
	Wind power penetration rate (proportion of output)	%	3	6	8	10
	Installed capacity of wind power	GW	22	55	104	279
	Proportion of installed capacity of wind power	%	6	10	13	15

As shown in Table 4.2.1, in 2025, 2030, 2040, and 2060, the installed capacity of wind power in AMS will be 15 GW, 27 GW, 53 GW, and 139 GW respectively, accounting for 4%, 5%, 7%, and 8% respectively.

For the Leap-forward Scenario, the installed capacity of wind power will be 22 GW, 55 GW, 104 GW, and 279 GW, respectively in 2025, 2030, 2040, and 2060, accounting for 6%, 10%, 13%, and 15% respectively.

To fully tap the development potential of wind power in ASEAN region, respond to climate change in a better way and accelerate the adjustment of energy mix, AMS can further enhance the development of wind power projects based on the existing development routes, and there is still room to improve the existing wind power development goals of AMS.

4.3 Analysis of LWSP Development

To further clarify the wind power development roadmap for AMS, the overall goals of medium to long-term for wind power development are decomposed into the specific countries in combination with the internal and external conditions for wind power development, such as endowment characteristics of wind energy resources, status and future trend of power grid development are, and policy support will be.

In terms of wind energy resources, Thailand, Vietnam, and the Philippines are endowed with the most abundant wind energy resources; Myanmar, Indonesia, Lao PDR, and Cambodia have a certain development potential of wind energy resources; while Malaysia, Singapore, and Brunei Darussalam are not rich in wind energy resources. The better the wind energy resources, the greater the national goal of wind power development.

In terms of the current power system status of AMS, the power grids of Singapore and Brunei Darussalam are relatively mature, and those of Thailand, Vietnam, and Malaysia are relatively perfect, which is conducive to the development of clean energy such as wind power. Considering the continuous implementation of APG program, it is expected that the power grids of Thailand, Vietnam, and Malaysia will develop rapidly in the future, which is beneficial to the development of clean energy projects. In the decomposition of wind power development goals, the better the power grid conditions and the better the development prospects are, the greater the goal of developing wind power will be.

In terms of power structure, the installed capacity of natural gas accounts for more than 75% in Brunei Darussalam and Singapore; for more than 40% in Thailand, Myanmar, and Malaysia; and for nearly 30% in Indonesia, which can provide relatively sufficient regulating capacity and create favourable conditions for large-scale development of wind power. Hydropower is the main source of installed capacity in Lao PDR, Myanmar, and Cambodia. The scale of wind power development can be expanded through the planning and development of hybrid energy system of hydropower, wind power, and PV power. Hydropower capacity accounts for more than 95% in Lao PDR, so the stability of output can be effectively improved through the hydrid energy system, to make up for power shortage caused by insufficient hydropower in dry season. The countries with good peak regulating capacity of power system and complementary demand of other power sources can undertake more wind power development.

In terms of the advantages of developing wind power, Thailand has a relatively high

electricity consumption. While other energies may not be sufficient to support the energy security in power sector, wind power could be an alternative option to improve the energy security in the future. Lao PDR is rich in hydropower resources, and its economy of hydropower development is better than other renewable energies. Therefore, the benefits of developing wind power are unclear in the short-term. Although Myanmar is rich in hydropower resources, the development of hydropower is suspended due to policy and environmental factors and wind power is a good choice for development. The countries with greater advantages in economy, resource endowment, and constraints of wind power, can undertake more wind power development.

In terms of supportive policies for wind power development, Thailand, Vietnam, and the Philippines have clearly defined the wind power development goals and implemented some supportive policies. Therefore, these countries can undertake more decomposed goals accordingly. Besides, Indonesia, Myanmar, Cambodia, and Lao PDR have started to promote the development of wind power projects and have also undertaken some wind power development tasks accordingly. However, according to the relevant policies of Malaysia, Singapore, and Brunei Darussalam, they do not make wind power a priority in future power development, and do not undertake or only undertake a few wind power development tasks. Based on the above factors, the overall wind power development goals for AMS at various planning level years under the APAEC Target Scenario as the development goal and the Leap-forward Scenario are decomposed as shown in Table 4. 3. 1. Overall, Thailand, Vietnam, the Philippines, and Myanmar have good development potential and conditions, which are the key to realizing the wind power development goals for AMS. In addition, Lao PDR, Indonesia, and Cambodia are also important to achieve the wind power development goals of AMS.

Table 4. 3. 1 Decomposition of recommended wind power generation goals for AMS in 2025, 2030, 2040 and 2060

Development scenario	Country	Energy output/TWh				Installed capacity/GW			
		2025	2030	2040	2060	2025	2030	2040	2060
APAEC target scenario	Brunei Darussalam	0	0	0	0	0	0	0	0
	Cambodia	1	2	5	14	1	1	2	6
	Indonesia	2	3	7	20	1	2	3	9
	Lao PDR	2	4	8	22	1	2	4	10
	Malaysia	0	0	0	0	0	0	0	0
	Myanmar	4	7	15	44	2	3	7	19
	The Philippines	4	7	16	45	2	4	7	20

Continued

Development scenario	Country	Energy output/TWh				Installed capacity/GW			
		2025	2030	2040	2060	2025	2030	2040	2060
APAEC target scenario	Singapore	0	0	0	0	0	0	0	0
	Thailand	10	19	42	119	5	9	19	52
	Vietnam	5	9	20	56	2	4	9	24
	Total	26	52	112	320	13	25	51	139
Leap-forward scenario	Brunei Darussalam	0	0	0	0	0	0	0	0
	Cambodia	3	5	10	28	2	2	5	12
	Indonesia	4	7	15	41	2	4	7	18
	Lao PDR	5	8	17	46	3	4	8	20
	Malaysia	0	0	0	1	0	0	0	0. 4
	Myanmar	10	16	33	90	5	8	15	39
	The Philippines	10	17	34	93	5	8	15	41
	Singapore	0	0	0	0	0	0	0	0
	Thailand	27	44	90	248	13	21	41	108
	Vietnam	13	21	42	116	6	10	19	50
	Total	72	119	240	665	36	57	109	289

4. 4 Key Development Regions and Projects

According to China's practical experiences, developing wind power projects in the form of wind power base is conducive to making unified planning and approval of wind power farms; realizing resource sharing, infrastructure, related facilities co-construction, joint operation and maintenance, and bundled generation delivery of wind power farms in the base. Moreover, the installed capacity of wind power can be effectively expanded, the overall cost of wind power development can be reduced, and the Research and Development (R&D) and manufacturing industry of related equipment can be facilitated. To avoid the influence arising from the variability of a single wind power source, the wind power base can implement hybrid energy system with solar energy, hydropower, coal, and natural gas, to improve the stability of wind power generation.

In accordance with the wind power development goals of the AMS, combining the characteristics of the spatial distribution of wind energy resources and ground surface cover, and taking into account the distribution of power grids, distance from load centres and other factors, 20 wind power bases/projects were selected in this study. The main information of each base/project is summarized as shown in Table 4. 4. 1.

Table 4.4.1 Statistical of planned wind power bases and projects in AMS

No.	Country	Province	Installed capacity /GW	LCOE/ (USD/ kWh)	Annual utilization hours/h	Capacity factor	Wind speed /(m/s)	Terrain	Total investment /(billion USD)	Land cover
1	Myanmar	Mandalay	1.46	0.055	2,373	0.27	6.0 - 7.0	Flat land	2.56	Dry farmland
2	Myanmar	Magway-Rakhine	1.22	0.05	2,675	0.31	7.0 - 7.5	Mountain land	2.14	Forest and shrubbery
3	Indonesia	Jawa Timur	0.16	0.057	2,452	0.28	5.0 - 7.0	Flat land	0.26	Mainly dry farmland, with a small amount of shrubbery and irrigated farmland
4	The Philippines	Negros Oriental	0.58	0.058	2,397	0.27	5.0 - 7.5	Flat land	0.96	Dry farmland and shrubbery
5	The Philippines	Iloilo	1.64	0.061	2,297	0.26	6.0 - 7.0	Flat land	2.71	Dry farmland and cultivated land
6	The Philippines	Nueva Ecija 1	0.35	0.042	3,307	0.38	7.0 - 7.5	Flat land	0.58	Dry farmland
7	The Philippines	Nueva Ecija 2	0.76	0.054	2,563	0.29	6.0 - 7.0	Flat land	1.25	Dry farmland
8	Vietnam	Ca Mau	2.8	0.056	2,103	0.24	5.0 - 6.5	Flat land	4.2	Grassland and dry farmland
9	Vietnam	Dak Lak	0.67	0.044	2,649	0.30	7.0 - 7.5	Flat land	1.01	Shrubbery
10	Vietnam	Gia Lai	1.07	0.047	2,499	0.29	6.8 - 7.5	Flat land	1.61	Shrubbery, with a small amount of dry farmland
11	Vietnam	Quang Binh	0.92	0.041	2,859	0.33	7.0 - 7.5	Flat land	1.38	Bare land and grassland
12	Lao PDR	Saravan-Champasak	3.92	0.059	2,194	0.25	5.0 - 6.0	Flat land	6.66	Shrubbery and dry farmland
13	Lao PDR	Svannakhet 1	2.06	0.054	2,397	0.27	6.0 - 7.0	Flat land	3.5	Dry farmland
14	Lao PDR	Svannakhet 2	3.51	0.057	2,271	0.26	6.0 - 7.0	Flat land	5.97	Dry farmland
15	Lao PDR	Svannakhet 3	1.11	0.05	2,590	0.30	7.0 - 7.5	Flat land	1.89	Dry farmland and irrigated farmland
16	Thailand	Surin	2.92	0.061	2,077	0.24	5.0 - 6.0	Flat land	4.96	Dry farmland
17	Thailand	Mukdahan-Kalasin	0.13	0.051	2,530	0.29	6.0 - 7.5	Flat land	0.22	Shrubbery
18	Thailand	Amnat Charoen-Ubon Ratchathani	1.23	0.051	2,471	0.28	6.0 - 7.0	Flat land	2.09	Dry farmland
19	Thailand	Roi Et-Kalasin	1.1	0.05	2,566	0.29	6.0 - 7.5	Flat land	1.87	Dry farmland and shrubbery
20	Thailand	Chaiyaphum	0.41	0.057	2,234	0.26	6.0 - 7.0	Flat land	0.7	Dry farmland

The wind power bases/projects preliminarily selected in this study are mainly distributed in the Philippines, Vietnam, Thailand, Lao PDR, and Myanmar, with a total installed capacity of 28.0 GW. Except Magway-Rakhine project in Myanmar, other wind power bases/projects are located in flatlands, and the land used is mainly bare land, grassland, and farmland. The capacity of the 20 wind power bases/projects ranges from 0.24 to 0.38, and the LCOE ranges from 0.041 USD/kWh to 0.061 USD/kWh, which is lower than the average LCOE level of 0.050 - 0.062 USD/kWh in the countries where they are located. The development of the selected bases/projects will drive the investment of 46.5 billion USD.

Box 4.2 Digital site selection method of wind power bases/projects

Box 4.2.1 Introduction

In this study, the wind power bases/projects were selected by the digital site selection method. With this method, it is first necessary to fully understand the time and spatial distribution of wind energy resources and their characteristics in the region, and preliminarily determine the suitable areas for the construction of wind farms. Then, a variety of restrictive factors should be considered comprehensively to eliminate unsuitable areas, such as the protected areas, rivers, and lakes, and to avoid using the land that should not be occupied too much in a centralized way, such as forest, cultivated land, and city. Second, different terrains suitable for development, such as flat and mountainous terrains, should be screened out by the geographical elevation data, and wind turbines should be automatically laid out in combination with WTGS selection. Finally, the main parameters such as the installed capacity of wind farm, should be analysed and calculated, and economic analysis should be made based on the grid connection conditions and external traffic conditions of the site, in order to acquire the estimation results of total investment and the average power cost per kWh.

Box 4.2.2 Layout of wind turbines

When the wind blows through the blades of the wind turbine, the wind turbine will absorb some of the wind energy, and the rotating blades will also increase the turbulent kinetic energy of wind, resulting in airflow distortion, turbulence, and a sudden decrease in wind speed, which are known as the wake effect. Nevertheless, if the distance between turbines is too small, which will result in the pitch variation due to change in wind direction, the influence of wake and turbulence intensity on adjacent wind turbines will increase, which will seriously endanger the safety operation of the wind turbines. Generally, the distance between wind turbines should be at least four times the rotor diameter. However, if the distance between wind turbines is too large and occupied area is large, the utilization rate of wind energy resources will be low, and the scale of bene-

fits will be insignificant, thereby affecting the economy of the wind farms. Based on a study conducted by Taleb and Hijleh on optimizing wind power generation in low wind speed region, wind energy production is highly affected by the layout of the wind turbines. The addition of smaller wind turbines between the original wind turbines, positioned at the front row facing the prevailing wind direction, would increase wind energy production. Nevertheless, the wind turbines spacing should be calculated carefully, as placing many wind turbines in close proximity would result in reduced energy production, while too wide spacing would not justify the land footprint. The appropriate wind turbine spacing should be selected with comprehensive consideration, in order to avoid the wake effect and improve the economy.

Box 4. 2. 2. 1 Layout on flat terrain

Under the stable terrestrial atmospheric conditions, the wind turbines in a flat wind farm should be arranged in an array, which can effectively reduce the wake effect. In the direction parallel to the prevailing wind direction, the wind turbines should generally maintain a distance of 5 – 9 times the rotor diameter, while in the direction perpendicular to the prevailing wind direction, the wind turbines should generally maintain a distance of 3 – 5 times the rotor diameter, and the front and rear wind turbines should be staggered in a plum blossom shape. The layout is shown in Figure 4. 4. 1.

Figure 4. 4. 1 Layout of wind farm in flat terrain

Box 4. 2. 2. 2　Layout on mountainous terrain

Under the stable terrestrial atmospheric conditions, wind turbines in mountainous wind farm should be mainly arranged in a single row according to the prevailing wind direction and terrain trend, which can effectively reduce the wake effect. The wind turbines perpendicular to the prevailing wind direction should maintain a distance of no less than two times the rotor diameter, while those in parallel to the prevailing wind direction should maintain a distance of no less than four times the rotor diameter. The layout is shown in Figure 4. 4. 2.

Figure 4. 4. 2　Layout of wind farm in mountain

Box 4. 2. 2. 3　Estimation of installed capacity of a wind farm

Based on the wind energy density map, the wind turbines are laid out according to different layout rules for plains and mountains within the region. After traversing all points that meet the layout requirements at the site, several layout schemes should be generated, from which the scheme with the greatest installed capacity should be selected as the final scheme. The number of wind turbines should be determined according to the layout, and the power value corresponding to a single wind turbine should be multiplied by the number of wind turbines arranged to acquire the installed capacity of the wind farm.

4.5 Research on Safeguard Measures for Flexibility of Power System

As wind power generation and PV power generation are variable and uncertain, the new energy power system needs to have quick adjustment capability and can dispatch various flexible power supply, in order to achieve the balance between supply and demand. For the power system with higher proportion of variable power source, such as wind power and PV power, it is required to have better regulation and balance ability. If the power system lacks flexibility, it will lead to the loss of a lot of clean power. According to IEA, from the perspective of power supply balance, the flexibility of power system is the ability to regulate and control the use of various energy resources and to operate safely, efficiently, cleanly, and economically in the case of the power system changes and uncertainties. To achieve the above goals, measures can be taken in three aspects, namely, power supply, power grid, and load.

4.5.1 Flexible power supply

Flexible power supply mainly plays a role of fast and deep peaking and operates complementarily with new energy generating sets. It can operate in the following models: in the peak season or period of wind power and solar power generation, the flexible power supply can make enough room for generation of new energy generating sets; while in the period with small outputs of wind power and solar power, the flexible power supply can supplement the insufficient space of new energy generating sets. Meanwhile, in all cases, the flexible power supply should provide necessary support for frequency, voltage, and rotational inertia to ensure safe operation of the system. At present, the flexible power supply mainly includes upgrade of thermal power plants, large-scale virtual synchronous machines, complement of multi-variety renewable energy resources, large-scale energy storage, etc.

(1) Upgrade of thermal power plants. The upgrade of thermal power plants is mainly embodied in two aspects, i. e. , improving the operation flexibility of generating sets, that is, the generating sets are required to have higher load change rate, higher load adjustment accuracy, and better primary frequency modulation performance; improving the flexibility of boiler fuel, that is, the boiler should be ensured to stable combustion and the generating sets still have good load adjustment performance in the case of different quality fuels used.

(2) Virtual synchronous machine. The traditional power systems are dominated by synchronous generators, and the inertia and damping of the generators play an important role in stable operation of the system. At present, most of new energy power generations are connected to the power grid through power electronic equipment, and these equipment with characteristics of low inertia and non-damping will impose a negative impact on stable operation of the power system. The virtual synchronous generator, referring to the operation of traditional synchronous generator, enables the new energy generating sets to have the inertia and damping characteristics

through reasonable control of grid connection device.

(3) Complement of multi-variety renewable energies. On the time scale, wind power, solar power, hydropower, and other renewable energy have power generation complementary characteristic to a certain extent. By virtue of this complementary characteristic, the power grid's ability to absorb new energy generation may be improved.

(4) Large-scale energy storage on power supply side. The large-scale energy storage on the power supply side is mainly in the form of traditional pumped storage power station and electrochemical energy storage emerging in recent years. With a history of more than 100 years, the pumped storage power station is relatively mature in technology, with the advantages of long service life and high overall efficiency. However, it has higher requirements on site selection and cannot be flexibly matched with other power stations like the electrochemical energy storage. The electrochemical energy storage has advantages of flexible configuration, short construction period and fast response, but its current economy cannot meet the requirements of large-scale promotion. AMS can configure flexible power sources according to their own conditions.

4.5.2 Flexible power grid

In order to improve the operation safety, stability, flexibility, and controllability, the flexible power grids must be able to quickly adjust and optimize power flow dynamically. The technological means for the flexible power grids mainly include Ultra-high Voltage (UHV) AC/DC transmission, flexible DC transmission, power flow control technology, etc. In addition, cloud computing, big data, artificial intelligence, Internet of Things, and other technologies are combined to realize the intelligence and digitization of power grid state perception, enterprise management, operation data control and so on, and improve the adaptability of power grid to new energy with high variability. It is also the development trend in the future.

4.5.3 Flexible load

According to IEA's research, the operating cost of power system can be reduced by 2%-11% by activating the demand side and storing energy, provided that the renewable energy development is not affected. The practice of flexible load is mainly embodied in the integrated energy system on the user side, virtual power plant, clean heating by "Internet Plus", "vehicle-network coordination" for electric vehicles, etc. According to the current practical experiences of China in the flexibility of power system, allocating electrochemical energy storage for wind farms in a certain proportion is a good way to improve the flexibility of power system and increase the consumption capability for new energy generation. However, after the investment in energy storage is included, the project income will be reduced significantly. Therefore, it is recommended to make allocation prudently in accordance with the specific new energy development and project situation.

Chapter 5

Experience and Successful Cases of LWSP Development

5.1 Important Issues and Experience

5.1.1 Power consumption

Wind power, solar power, and other new energy sources are highly variable and do not have the regulation ability. When the output does not match the demand, wind power and solar power might have to be curtailed. In China, for example, during the "12th Five-Year Plan" period, the installed capacity of wind power showed explosive growth, especially in Northwest China, North China, and Northeast China with abundant wind power resources. The construction of supporting power grids does not match the installation of WTGS, the local power consumption is limited, and the outgoing power transmission is insufficient. As a result, wind power is curtailed seriously. According to statistics, China's wind power curtailment rate reached 13% from 2011 to 2015, and 17.1%, the worst in 2012. The high rate of wind power curtailment has discouraged wind power investment. After 2016, the Chinese government has taken measures to promote wind power consumption and wind power curtailment has gradually been alleviated.

(1) Guaranteed purchase. In 2016, the National Development and Reform Commission (NDRC) of China officially issued the *Measures for Management of Full Guaranteed Purchase of Renewable Power Generation*, which divides the annual output of renewable grid-connected power projects into a guaranteed purchase share and a market-traded share. According to the measures, the guaranteed purchase share is purchased fully at the benchmark FiT by prioritizing the annual power generation plan and signing a priority power generation contract (physical contract or contract for difference) with the power grid company; and renewable power producers are encouraged to give full play to the advantage of low marginal cost of renewable power to achieve priority power generation and promote full-capacity power generation beyond the guaranteed purchase share.

(2) Construction of UHV power transmission lines. In 2014, the NEA of China proposed to accelerate the construction of 12 key power transmission lines in the air pollution prevention and control action plan, among which there are nine UHV power transmission lines. Eight of the nine UHV power transmission lines undertaken by the State Grid Corporation of China, together with Jiuquan-Hunan UHV DC Power Transmission Line and Jarud-Qingzhou UHV DC Power Transmission Line, have increased the cross-regional power transmission capacity in East-central China by 80 GW, which is equivalent to reducing coal consumption by 180 million tons, CO_2 emissions by 320 million tons, SO_2 emissions by 880,000 tons, and NOx emissions by 940,000 tons per year, if they are all used to deliver clean energy. This is also conducive to cross-regional clean energy consumption.

(3) Risk monitoring for power investment. In view of the serious curtailment in some areas around 2015, NEA began to issue a warning for national wind power investment in 2016. The risk levels are "red", "orange", and "green" from high to low, and the monitoring target is the next year after risk issuance. For the "red" areas, NEA will not set the annual development and construction target in the year of risk issuance, local authorities will suspend the approval of new wind power projects, wind power development enterprises are advised to make careful decisions on the construction of wind power projects, and power grid companies no longer deal with new grid connection procedures. For the "orange" areas, which are normal, local governments and enterprises can reasonably advance the development, investment, and construction of wind power projects according to market conditions.

The risk monitoring mechanism is mainly to guide rational investment in wind power and promote the sustainable and healthy development of the wind power industry. Jilin, Heilongjiang, Gansu, Ningxia, and Xinjiang were "red" regions in 2016 when the first warning was issued, and China had no "red" regions in 2020. The mechanism has effectively discouraged investment in wind power projects in the red regions, so that the curtailment of wind power has been alleviated to a certain extent.

(4) Peaking by other power sources. To solve the problem of wind power and other clean energy consumption, China has accelerated the construction of pumped storage power stations, large hydropower stations and other peaking power sources, upgraded hydropower stations and unit peaking, improved the utilization of existing transmission lines, and improved the flexibility of thermal power plant through decoupling or technical transformation of generation units to reduce the minimum technical output of units from the 50% to 40%-20%. At the same time, to increase the peaking capacity of the grid, China also encourages the development of electrification, and the construction of energy storage equipment is required in some regions to support the development of wind power and other clean energy power.

5.1.2 Business model

A mature business model can help attract investment, reduce costs, and effectively promote the development of the wind power industry. Many AMS are still in the initial stage of wind power development and need to explore a business model suitable for their conditions. From the existing international experience, the following models can be referred to:

(1) Local government-led wind power development. The relevant preliminary work is partially or fully completed mainly by the government, and then development enterprises are selected for specific sites. As a result, the government obtains shares of the project or part of the revenue. Some governments, directly or through a governmental platform, commissions a design and consulting company to complete wind power development planning and clear the development plan for each wind farm. The real case of this model which is currently available is Public-Private Partnership (PPP), which allows private companies to invest in vital project infrastructures including the power projects. This mode has the advantage of coordinating the regional wind energy resources, construction conditions, consumption environment and development restrictions, significantly reducing project uncertainties, promoting cooperation among different developers in the construction of step-up substations, roads, and other infrastructure by means of regional planning, and reducing development costs.

(2) Hybrid power development. As the output of wind power is variable, some wind power enterprises have made attempts at hybrid power generation, mainly including the complementation of new energy micro-grids, wind power, solar power and hydropower, integrated energy system solutions and "wind power+". Hybrid power generation projects are mostly in the demonstration stage and are also supported by local governments and power grid companies. A solar-wind hybrid design method, proposed by Hongxing et al., can be used as a reference in designing a hybrid solar-wind power generation system. The proposed design of hybrid solar-wind system has been applied to supply power on a remote island along south-east coast of China.

(3) Project exit mechanism. As some enterprises attach importance to project contracting but implement contracts inactively, local governments, in addition to making clearer requirements on project development schedule, can negotiate developers to own their wind measurement equipment and related data in case that they exit from project development midway. Considering that the developers have taken local development resources but not utilized them properly, the requirement is reasonable. Because of the existing data, it can significantly improve the development timeframe of new developers.

(4) EPC for the preliminary stage of the project. The site selection, wind measurement, and approval of a project are often completed independently by the owner. Because some small-scale investors or new entrants to the wind power industry have insufficient

experience, technical strength, and human resources, they may act slowly in the early advancement of projects. Therefore, investors can entrust other experienced design and consulting institutes to complete the relevant work in the form of early general contracting, thus optimizing the efficiency of project promotion.

(5) Financing+EPC (F+EPC). For the problem of difficult and expensive financing in some projects, developers can integrate financing, engineering, procurement, and construction trough the F+EPC mode. The F+EPC mode can solve the problem of limited financing channels and expensive capital cost in the project site, and at the same time, give full play to the advantages of the project and the good reputation of the contractor, and make full use of the contractor's own financing channels to reduce project financing cost. The contractor brings capital to the wind farm and wins the contract for the project, while the project can be successfully promoted with a win-win result.

(6) Wind power base. The base development mode can be implemented for large-scale wind power projects. The government can allocate the wind power resources through competitive bidding. In the process of base development, the government can select some of the contract winners on a merit basis to unify the construction of public roads, water supply and drainage, power supply, and other infrastructure within the base, while encouraging the preferred enterprises to bundle different projects into one big project to unify the early work and handle the relevant supporting documents, with other contract winners sharing the costs according to their contractual share.

5.1.3 Policies

From the existing international experience, the relevant supporting policies for renewable power mainly include supportive policies and constraining policies. Supportive policies mainly include tariff and subsidy policies, tax incentives, and mandatory portfolio system. Tariff and subsidy policies and tax incentives are important supporting policies in China and the United States. In other countries, market-based mechanisms such as auctions, tenders, and green certificates are the main drivers of wind power development. Constraining policies mainly include environmental policies and investment access systems. Usually, countries will choose different policy combinations according to their own conditions. For example, China and the United States mainly adopt tariff and subsidy policies and tax incentives, while some European countries promote the development of renewable energy through bidding, green certificates, and other methods.

5.1.3.1 Power planning and consumption policies

In 2015, China proposed to coordinate the planning for local consumption and power transmission of wind power bases. First, multiple measures should be taken to accelerate the local consumption of wind power, while focusing on the construction of wind power bases and the use of cross-provincial or cross-regional transmission lines to expand the allocation of wind power resources. Inner Mongolia, Xinjiang, Ningxia, Gansu, Shanxi,

Shaanxi, and other provinces (regions) were required to accelerate the construction of the supporting wind power bases for the planned cross-regional and cross-provincial transmission lines according to the transmission line plan, the deployment of air pollution prevention and control work and include the bases into the wind power development plan for Northwest China, North China, and Northeast China during the "13th Five - Year Plan" period.

In 2019, China proposed to establish a sound mechanism for guaranteeing the consumption of renewable power, and defined responsibilities for renewable power consumption in each provincial administrative region. The responsibility would be comprehensively monitored and formally assessed from 2020.

5.1.3.2 Tariff policies

Tariff and subsidy policies are one of the key measures to support the stable development of the renewable power industry. According to the national conditions and different stages of renewable power development, the tariff policies for renewable power include FiT, Contracts for Difference (CfD), net metering, tendering, etc.

(1) Feed-in Tariff (FiT). In the initial stage of renewable power development, most countries adopt the FiT policy, that is, the FiT is specified according to the cost of electricity generation of various types of renewable power, and the power grid companies pay the power producers according to this standard. The difference between the FiT and the market tariff is made up by subsidies. By the end of 2019, 113 countries or territories worldwide had implemented FiTs for renewable power.

(2) Contracts for Difference (CfD). In countries with market-based electricity trading, parties enter long-term CfD and agree on a fixed "strike price" for the duration of the contract to obtain a stable level of tariff. The CfD is executed in such a way that the renewable power producer in the CfD sells electricity through the power market and then receives payment for the difference between the market tariff and the strike price. When the market tariff is higher than the strike price, the power producer needs to return the price difference; when the market tariff is lower than the strike price, the power producer can get the price difference compensation. This tariff mechanism provides effective, stable, and long-term support for renewable power, and at the same time gives investors greater certainty in terms of project returns, thus reducing the financing costs and risks of projects. A representative country that has implemented a CfD policy is the United Kingdom.

(3) Net metering tariff. This approach allows consumers with renewable power generation facilities to deduct a portion of their electricity from their bills based on the amount of electricity delivered to the grid, i. e. , only "net consumption" is calculated. Net metering tariffs are typically used for small-scale generation facilities on the customer side, such as wind power, solar power, and home fuel cells.

(4) Tendering. With the market-based development of renewable power, determi-

ning tariff through tendering has become the international trend of pricing in more and more countries. From the tendering results in some countries, this mechanism has played a big role in lowering renewable power cost and tariff, gradually got rid of subsidy dependence and moved into the market stage. The tendering mechanism has had an important impact on the decline in wind power cost and tariff.

5.1.3.3 Taxation policy

Taxation policy is an important tool for a country to implement macro-control. For the renewable power industry, targeted provision of preferential taxes and incentives can attract more project investment and expand the scale of renewable power development. From the global situation of countries that have carried out renewable power tax incentives, they mainly focus on tax relief and credits for equipment production, investment, environmental protection, and other aspects.

The United States, for example, has the best renewable power tax incentives, and the government supports the development of non-fossil energy by providing tax credits for wind power, solar power, and biomass power mainly including production tax credit (PTC) and investment tax credit (ITC). The PTC has been in place since 1992, and credits are available for electricity produced by qualified renewable power equipment and delivered to power grids (on a kWh basis). The credits are generally available for the first 10 years after the equipment is placed in service. The PTC has been instrumental in the growth and development of renewable power in the United States, particularly wind power, and is one of the key drivers of wind power project development in the United States.

China also provides tax incentives for the wind power industry, mainly including preferential Value-added tax (VAT) and income tax policies. The preferential VAT policies mean that the value-added input tax on fixed assets in wind power investments can be offset against value-added output tax on income from wind power sales, in addition to a 50% preferential policy of immediate VAT refund for wind power. The preferential income tax policies include the "three exemptions and three halves" policy and the "Western Development" policy. The "three exemptions and three halves" policy means that wind power investment enterprises are exempt from corporate income tax from the first to the third year of operation and exempt from 50% of corporate income tax from the fourth to the sixth year of operation. The "Western Development" policy means that corporate income tax is reduced from 25% to 15% for wind power farms built in western China.

5.1.3.4 Mandatory portfolio system

The mandatory portfolio system for renewable power, as an incentive policy, refers to legally specify the proportion of renewable power in the total power generation, and require power grid companies to purchase the renewable power. The power grid companies must bear the legal responsibility if failing to meet the portfolio requirements. The policy is implemented in conjunction with a renewable power certificate system. Currently, the

mandatory renewable portfolio system is widely used in the United States, the United Kingdom, Australia, and Japan.

The United States is the first country to implement the RPS, and it is also the country that has produced good results. The RPS has been in place since the 1990s in all states of the United States, with the power suppliers providing a minimum share of renewable power (usually including wind power, solar power, biomass power, geothermal power, etc.) according to each state's program. The power suppliers can fulfil their portfolio mandate by operating renewable power generation facilities or by purchasing renewable energy credits (RECs). By 2019, 29 states and Washington D. C. have implemented the RPS in the United States, stipulating specific renewable power targets according to their resource conditions and development directions, leaving the specific production and trading to the market, with the government setting the relevant standards and the provisions of trading, verification, and penalty. The RPS has facilitated the shift from relying on financial support in the past to government-regulated market mechanisms, thus creating conditions for large-scale development of renewable power.

5.1.3.5 Environmental impact policy

To reduce the negative impact of wind farm construction and operation on the environment, environmental impact policies need to be implemented. China, for example, formulated the *Interim Measures for the Management of Wind Farm Construction Land and Environmental Protection* in August 2005, which gives clear requirements and approval process for both wind power construction land and environmental protection. Wind farm construction land should be based on the principle of economical and intensive use of land, using unused land as much as possible, occupying less or no arable land, and avoiding areas that require special protection as approved by government departments at or above the provincial level. The environmental impact assessment system is implemented for wind farm construction projects. The developer should obtain the approval of environmental impact assessment before application for project construction. China has also implemented a protection system for public welfare forests, ecological red lines, cultural relics, water conservation areas, important tourist areas, and wetlands, so that wind farms are as far away from protected areas as possible and the impact of wind power development on the environment is minimized.

5.1.4 Development costs

According to a report by IRENA on the cost of renewable power, the cost of renewable power generation has fallen dramatically over the past decade, driven by technological advances, economies of scale, increasing competition in the supply chain, and growing developer experience. According to cost data collected by IRENA from 17,000 projects in 2019, the costs of onshore and offshore wind power have fallen by 39% and 29%, respectively, since 2010.

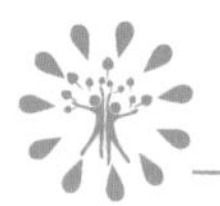

According to the Global Wind Energy Council (GWEC), the cost of wind power has fallen by 60% due to economies of scale and 40% due to technological advances. It is projected that onshore wind costs could fall by about 15% by 2030 and about 20% by 2050 compared to those in 2020.

In addition to wind power development costs, there is also room for project O&M costs to decline. With the development of smart turbines, remote control technology, UAV patrol, and other technologies, the efficiency and labour costs of wind farm O&M will further decline, while the equipment failure rate will be further reduced as wind power equipment further matures. There is also room for improvement for the transparency of land licensing processes and procedures which should ease the process for developers during the land acquisition stage. This process may add to the cost of licensing fees, therefore, having safeguards or transparency during the land licensing process can reduce these additional costs.

With the decline in wind power costs, not only wind power parity becomes possible, the economic benefits of low-speed wind farm development will also continue to improve.

5.2 Successful Experience

5.2.1 Panyang Wind Farm

(1) Project overview: Panyang Wind Farm is located in Gutian County, Fujian Province, China, with an installed capacity of 48 MW and 24 units of WD115 - 2000 low-speed wind turbines manufactured by Windey. The project site has a high altitude, with the turbines at the altitude between 1,200 m and 1,487 m. The annual average wind speed at the height of 85 m is 5.74 m/s, and the farm is a typical coastal mountainous low-speed wind farm (Figure 5.2.1).

Figure 5.2.1 Panyang Wind Farm

(2) Project features: Due to the specific geographical location and altitude, thunderstorms and freezing are the two biggest problems for Panyang Wind Farm. In order to guarantee the absolute safety of the unit operation, the project refers to the design standards for lightning protection of projects in plateau areas and adopts reliable lightning protection measures for key equipment such as blades, anemometers, nacelles, control cabinets, and converters. The treatment of anti-freezing is based on the anti-freezing experience of wind farms in southern China, and a highly targeted anti-freezing design has been adopted.

(3) Project operation: As the first low-speed wind farm in Fujian's coastal mountains, Panyang Wind Farm has been operating well since it was put into operation, and the wind turbines have withstood the test of many strong thunderstorms. On January 24, 2016, the project withstood the test of the "giant" cold wave, all units operated normally, and the power generation on that day exceeded 630 MWh with the availability of all units reaching 100%. From the Commercial Operation Date (COD) to December 2020, the total power generation of the 24 units reached 574.5 GWh, with an average annual equivalent power generation hours of 2,710.

(4) Demonstration significance: In Panyang Wind Farm, the Windey wind turbines were customized for the specific wind farm environments such as thunderstorms and typhoons to ensure safe and efficient operation of the turbines and increase the turbine availability, thus enable low-speed wind farms to generate much power. The experience of Panyang Wind Farm can provide a reference for wind power development in AMS with special climate conditions of thunderstorms and typhoons.

5.2.2 Jindi Wind Farm

(1) Project overview: Jindi Wind Farm in Hua County, Henan Province, China, has a total planned capacity of 80 MW, a site area of 11 km^2 and an average altitude of about 50 m. The site is located a plain area mainly with crops as the vegetation. The project has a hub height of 100 m, and the annual average wind speed at the hub height is 5.26 m/s. 40 Goldwind GW2.0 (S) MW smart wind turbines are installed. It is a typical low-wind-speed wind farm in plains (Figure 5.2.2).

(2) Project features: Jindi Wind Farm in Huaxian County is on the north bank of the Jindi River, making it a beautiful scene along the river. The wind farm was designed and constructed under the guidance of the local planning, environmental protection, transportation, water conservancy, township, and other departments, combines with the high standard demonstration fields and the Jindi River landscape belt, and was successfully built as a wind farm demonstration project from multiple angles and multi-dimensions. The wind turbines are mainly installed in agricultural land. Through the use of different tower foundations, local conditions and the agricultural land, wind turbines are installed without changing the use, affecting the function, and increasing the footprint, and the

Figure 5. 2. 2 Jindi Wind Farm

project can be replicated elsewhere. Jindi Wind Farm has further lowered the wind speed for power development to 5. 26 m/s.

(3) Project operation: The project was put into operation in April 2017, and as of April 2021, the wind farm is in good operating condition, with a unit availability of 98%, a total power generation of 53,328 MWh and annual equivalent utilization hours of about 2,020.

(4) Demonstration significance: Jindi Wind Farm is a wind power project integrating with farmland and agricultural land, making farmland less affected through specific land plans. The wind farm has a wind speed of only 5. 26 m/s, and the technology in 2017 can make the wind farm utilization hours reach more than 2,000. The current wind energy utilization efficiency and power generation will be higher. Vietnam and Thailand have large farmland areas with a wind speed of 5. 2 m/s or higher, and wind farm development can be integrated with farmland and corresponding facilities to effectively improve the development potential of wind power projects by learning from the Jindi Wind Farm.

5. 2. 3 Dongbatou Distributed Wind Farm

(1) Project overview: Dongbatou Distributed Wind Farm is located in a plain terrain in Lankao County, Henan Province, China, with an altitude between 65 m to 75 m and an annual average wind speed between 6. 0 m/s and 6. 1 m/s. The wind farm is a distributed project with a total installed capacity of 11 MW, five 2. 2 MW wind turbines, and a total investment of about RMB 100 million (Figure 5. 2. 3).

(2) Project features: Dongbatou Wind Farm in Henan Province started its feasibility study, preliminary wind measurement and wind resource demonstration in 2017. The

Figure 5. 2. 3 Dongbatou Wind Farm

demonstration shows that the project has a large wind shear and increasing the tower barrel height can significantly increase the revenue. As a result, China's first 140-meter-tall tower barrel was installed in the project.

The project is located in the central plain of China, and the project site is all farmland, mainly growing wheat, corn, and other crops. The land occupied by the installation platform and road during construction was restored afterwards and did not affect the growth of the original crops. In addition, the project uses a single blade hoisting solution, which further reduces the temporary land area during the construction period compared with the conventional wind turbine hoisting. The project is a distributed wind power project with local power connection and consumption. The wind turbines are connected to a 110 kV switchyard through 35 kV box-type step-up substations. The 110 kV switchyard is pre-assembled to further reduce land used and construction period, thereby reducing overall construction costs.

(3) Project operation: In January 2017, the project owner signed the development agreement with the Government of Lankao County. In June 2017, the project was put into operation only six months after project approval. Up to now, the project has operated in good condition for about four years. For the whole year from January 2020 to December 2020, a total of 24. 54 GWh of electricity was generated, with annual equivalent utilization hours of about 2,230.

(4) Demonstration significance: The project creatively uses the 140 m tall flexible steel tower barrel solution. During construction and operation, the industry-leading single-blade hoisting process, pre-assembled switching station and other advanced processes and technologies were used and greatly shortened the construction period and costs. The vortex excitation problem during the hoisting of the tall tower barrels was overcome with

spoiler bars, cable ropes, and other systems. During operation, control means such as active damping injection, rotor imbalance compensation and dynamic rapid traversal are also innovatively used to ensure the safety and stability of tower barrel operation. The project can provide a reference for the similar low-speed wind farms in high shear areas in ASEAN. The project is a distributed wind farm and is connected to the local power grid for consumption, providing demonstration significance for the development of small-scale wind power projects in ASEAN. In addition, the wind energy utilization of the wind farm is integrated with local agricultural production, to achieve a win-win situation of power generation and agricultural production, efficient use of local land and space, and maximize the benefits of land.

5.2.4 50 MW Chinh Thang Wind Power Project in Vietnam

(1) Project overview: 50 MW Chinh Thang Wind Power Project is located in the middle of NinhThuan Province, Vietnam. The wind farm is surrounded by mountains on the north, west and south, and about 12 km from the coastline on the east, with a generally flat terrain similar to the bottom of an open basin and an altitude between 14 m and 25 m. The surface is mostly fields, cultivated with cash crops. There is a 50 m wide river running through the middle of the wind farm.

Within 200 km of the project site, there is no strong typhoon (wind speed over 37.5 m/s) in the past 20 years, and the common typhoon intensity is 7 - 10 (wind speed at 16 - 25 m/s), so the site is less threatened by typhoon. The wind farm has an annual average wind speed of 6.29 m/s and is equipped with 13 WD147 - 3000 wind turbines and three WD147 - 3600 wind turbines with a hub height of 125 m. According to the estimation, the annual operating hours can reach 2,830, and the annual power generation is about 118,524 MWh (Figure 5.2.4).

Figure 5.2.4 50 MW Chinh Thang Wind Power Project in Vietnam

(2) Project operation: Two wind turbines have been installed as of January 21, 2021, and the project is scheduled to be put into operation in the second quarter of 2021.

(3) Demonstration significance: The project is an important low-speed wind power project in Vietnam, and its successful implementation will provide an important reference for the application of low-speed wind farms in ASEAN.

Chapter 6

Suggestions on LWSP Development in ASEAN

6.1 To Set a Clear Goal and Adopt Incentive Policies

It is important to integrate the country's wind energy resource conditions, fully exploit the potential of wind power in achieving emission reduction and energy transition, include the goals of wind power development into the overall goal of renewable power development, set the medium and long-term development targets of wind power, and actively promote the implementation of the incentive policies for wind power development.

In terms of FiT policy, in countries where the wind power industry is still in its infancy, reasonable FiT policies should be introduced. While in countries where wind power development has reached a certain scale, an open bidding model can be adopted to determine FiT and select investors to reduce the cost of electricity.

In terms of power consumption, countries and regions which have a reliable power system can promote full absorption of wind power generation. Regions with a weak power system or the isolated areas can adopt a PPA to guarantee the purchase of wind power.

In terms of administrative approval, special approval policies should be formulated for wind power and other new energy to simplify the approval process and shorten the approval time. The focus should be placed on the approval matters related to the public interest such as policies for new energy and environmental protection. Independent decision-making and risk sharing by enterprises should not be included in the conditions for approval, so as to minimize administrative intervention in new energy projects and create a favourable policy environment for the development of new energy in AMS.

6.2 Layout Optimization to Promote the Development of Wind Power Projects in an Orderly Manner

Wind power development should be accelerated in Thailand, Vietnam, the Philippines, Myanmar, and other countries with relatively good wind power development potential and conditions. Lao PDR, Indonesia, Cambodia, and other countries with certain wind power development potential can become an important complementary force to achieve the goal of ASEAN wind power development. Malaysia, Singapore, Brunei Darussalam, and other countries with limited resources can introduce wind power technology at the appropriate time and develop micro and distributed wind power projects to facilitate energy transition, depending on the development of wind power technology in the future.

To the key areas of wind power development, it is recommended that priority be given to areas close to load centres with better wind energy resources and better power grid and transportation conditions, with wind power bases or projects laid out mainly in central and southern Vietnam, southern Lao PDR, central and eastern Thailand, central Myanmar, central and northern the Philippines, etc. Consideration can be given to the integrated development of wind power with other renewable power sources such as solar power and hydropower, to avoid the volatility of wind power and improve the quality of wind power sources through hybrid power generation.

6.3 Development of ASEAN Wind Power Technical Standards According to Local Conditions

On the basis of combining the national conditions and resource conditions of each country and referring to the existing international standards, the wind power construction standards should be improved in the ASEAN region, so that the standards can not only ensure the quality of construction, speed up the construction progress, but also well control the project cost.

In terms of operation and dispatching, standards should be unified in terms of communications methods, transmission channels, and information transmission between wind farms and power system dispatching agencies, to realize real-time communications of telemetry signals, telematics signals, remote control signals, remote dispatch signals, and signals of other safety and automation devices among wind farms in different regions, thereby facilitating unified and optimal dispatching of different wind farms.

Standards should be prepared for wind turbine selection, giving priority to grid-friendly turbines, encouraging wind farms to participate in the regulation of the local power system with their peaking capacity, frequency, and voltage tolerance, or connect to the power grid together with solar power, hydropower, coal power, gas power, and other power sources, thereby reducing the volatility of wind power output.

In terms of wind turbine adaptability, according to the characteristics of low-speed wind energy resources in AMS, guidelines should be developed for low-speed wind farms, and priority should be given to the selection of wind turbines with large capacity, large rotor diameter, and high-power generation efficiency, so that wind turbines can fully adapt to local wind conditions. Wind turbines should have anti-typhoon characteristics in typhoon areas. Some wind farms with high wind shear should be provided with different tall-tower solutions. Considering that some wind farms may be close to residential areas, wind turbines should also have a low noise level with noise reduction measures.

6.4 Upgrading Measures for Flexibility of Power System

As wind power and solar power generation are causing the variability and hence supply uncertainty to the power system, the new energy power system needs to have a quick adjustment capability and the ability to dispatch various flexible power supply. One of the most effective measures to provide the flexibility of power system is through transmission interconnection. It allows the mobilisation of flexible resources (fast-dispatchable power plants) to mitigate the variability among the adjacent connected power areas.

Some countries in ASEAN are still in doubt to increase their penetration of variable renewable energy (VRE) due to this variability problem. Therefore, to support the development of wind power, upgrading the transmission grid to increase flexibility support is essential. This is also to support the new context of export-import electricity or power trade in ASEAN, especially to trade the electricity coming from VRE resources. The continuous implementation of APG program could be beneficial to the development of clean energy projects. As such, two factors are needed in realizing successful wind power development. First, robust interconnection and second, wind potential in the region.

6.5 Innovation and Exploration of Diversified Financing Models

It is important to encourage investment in wind power projects by improving investment environment and adopting innovative business cooperation models. A favourable environment should be created for wind power investment to promote LWSP projects through a variety of business models.

The investment environment should be further optimized for new energy projects, focusing on guarantees or commitments on grid connection, power consumption, land acquisition, and construction approval, to reduce investors' doubts and stabilize project revenue expectations.

In terms of investment and financing, communication should be conducted with multilateral institutions such as the World Bank, Asian Development Bank (ADB), and Asian Infrastructure Investment Bank (AIIB), as well as policy banks in Japan, South Korea, China, Europe, and the United States, to increase the support of major financial institutions for new energy (especially wind power) projects in the ASEAN region and reduce the financing costs of wind power projects.

In terms of business models, the mature international business models can be referred to for wind power development. By absorbing the advantages of different business models, combined with the local conditions, a business model for wind power development suitable for ASEAN's conditions can be created.

6.6 Win-win International Cooperation on Wind Power

Mature wind power technologies should be fully utilized, especially the latest technology suitable for ASEAN's low wind speed region and unique climate conditions. Technology-related capacity building should be enhanced, with a focus on local training, to improve ASEAN's technical capabilities in wind power development.

Wind power industrial chains should be introduced into ASEAN to localize equipment manufacturing. ASEAN can even take advantage of its geographical position to become a global centre for wind power equipment manufacturing and export, to reduce the cost of wind power development.

Efforts must be made to seize opportunities presented by RCEP to pursue multilateral and bilateral cooperation, collaborating with neighbouring countries and developed countries. Multilateral and bilateral mechanisms such as the East Asia Summit, ASEAN+3, ASEAN+1, must be fully utilized to make and coordinate high-level policies, build consensus on development, and jointly promote wind power development in ASEAN.

As a way forward, developing demonstration projects would be crucial as an effort to further develop low wind speed projects in ASEAN. In addition, the pilot project will also be a significant way to increase China-ASEAN cooperation practices. Therefore, further collaborative research needs to be carried out along with a series of capacity building for the development of low wind speed turbines, in order to realize the possibility of commercialization of pilot projects in several areas.

References

[1] ASEAN Centre for Energy, China Renewable Energy Engineering Institute. The Present Layout and the Prospect of ASEAN Power Interconnection Projects [R]. Beijing: CREEI, 2019.

[2] ASEAN Centre for Energy, China Renewable Energy Engineering Institute. Practical Experience and Prospects for Electricity Accessibility in ASEAN [R]. Beijing: CREEI, 2020.

[3] ASEAN Centre for Energy, China Renewable Energy Engineering Institute. The Roadmap for Innovative Scale Development of Solar PV in AMS and the Suggested Applications [R]. Beijing: CREEI, 2020.

[4] ASEAN Centre for Energy. ASEAN Plan of Action for Energy Cooperation (APAEC) 2016 - 2025 Phase Ⅰ: 2016 - 2020 [R]. Jakarta: ACE, 2015.

[5] ASEAN Centre for Energy. ASEAN Plan of Action for Energy Cooperation (APAEC) 2016 - 2025 Phase Ⅱ: 2021 - 2025 [R]. Jakarta: ACE, 2020.

[6] ASEAN Centre for Energy. The 6th ASEAN Energy Outlook 2017 - 2040 [R]. Jakarta: ACE, 2020.

[7] ASEAN Centre for Energy. The Paris Agreement and the Energy Policies of the ASEAN Member States: Policy Brief [R]. Jakarta: ACE, 2020.

[8] ASEAN Centre for Energy. Draft ASEAN Power Report 2021 [R]. Jakarta: ACE, 2021.

[9] ASEAN Centre for Energy. ASEAN Climate Change Energy Project (ACCEPT) NDCs compilation [EB/OL]. (2018 - 01 - 01) [2021 - 06 - 30]. https://accept.aseanenergy.org/country/.

[10] ASEAN Centre for Energy. ASEAN Energy Database System (AEDS) [EB/OL]. (2018 - 01 - 01) [2021 - 06 - 30]. https://aeds.aseanenergy.org/.

[11] BARNES R H, MOROZOV E V, SHANKAR K. Improved methodology for design of low wind speed specific wind turbine blades [J]. Composite Structures, 2015, 119: 677 - 684.

[12] BloombergNEF. 2H 2020 LCOE Update [R]. Shanghai: BloombergNEF, 2021.

[13] CHANG Y, HAN P. Harnessing Wind Energy Potential in ASEAN: Modelling and Policy Implications [J]. Sustainability, 2021, 13 (8): 4279.

[14] Department of Alternative Energy Development and Efficiency, Ministry of Energy of Thailand. Alternative Energy Development Plan 2018 - 2037 (AEDP 2018 - 2037) [R]. Bangkok: Ministry of Energy of Thailand, 2018.

[15] Enterprize Energy. Thang Long Offshore Wind Power Project [EB/OL]. (2018 - 09 - 01) [2021 - 06 - 30]. https://enterprizeenergy.com/projects/thang-long-offshore-wind-power-project-3.html.

[16] Global Wind Energy Council. Global Wind Report 2020 [R]. Brussels: GWEC, 2021.

[17] YANG H, ZHOU W, Lou C. Optimal design and techno-economic analysis of a hybrid solar-wind power generation system [J]. Applied Energy, 2009, 86 (2): 163 - 169.

[18] International Energy Agency. Guide to Balancing Challenges by Variable Renewable Energy [R]. Paris: IEA, 2011.

[19] International Energy Agency. Status of Power System Transformation 2019 [R]. Paris: IEA, 2019.

[20] International Energy Agency. World Energy Outlook 2020 [R]. Paris: IEA, 2020.

[21] Imperial College London. The flexibility of gas: what is it worth? [R]. London: Imperial College London, 2020.

[22] International Electrotechnical Commission. International Standard of Wind Turbines IEC 61400 - 1 [R]. Geneva: IEC, 2005.

[23] International Renewable Energy Agency. Estimating the Renewable Energy Potential in Africa: A GIS - based approach [R]. Abu Dhabi: IRENA, 2014.

[24] International Renewable Energy Agency. Renewable Energy Statistics 2020 [R]. Abu Dhabi: IRENA, 2020.

[25] International Renewable Energy Agency. Renewable Power Generation Costs in 2019 [R]. Abu Dhabi: IRENA, 2020.

[26] Ministry of Commerce of China. Country (Region) Guide for Foreign Investment Cooperation: ASEAN (2020 Edition) [R]. Beijing: Ministry of Commerce of China, 2021.

[27] Ministry of Industry and Trade of Vietnam. 55 - NQ/TW Resolution on Orientations of the Vietnam's National Energy Development Strategy to 2030 and outlook to 2045 [R]. Hanoi: MOIT, 2020.

[28] Ministry of Industry and Trade of Vietnam. Draft Vietnam PDP8 [R]. Hanoi: MOIT, 2021.

[29] Ministry of Energy and Mineral Resources Republic of Indonesia. National Energy Plan (RUEN) [R]. Jakarta: MEMR, 2020.

[30] National Renewable Energy Laboratory. Exploring Renewable Energy Opportunities in Select Southeast Asian Countries [R]. Golden: NREL, 2019.

[31] PAGERPOWER, [EB/OL]. Shadow Flicker Assessments for Wind Tuebines [EB/OL]. (2016 - 03 - 18) [2021 - 06 - 30]. https: //www. pagerpower. com/news/shadow - flicker/.

[32] PLN 2018 Annual report [R]. Yogyakarta: PLN, 2019.

[33] Protected Planet. Discover the world's protected areas [EB/OL]. (2014 - 01 - 01) [2021 - 06 - 30]. https: //www. protectedplanet. net/en.

[34] Renewables 2020 Global Status Report [R]. Paris: REN21, 2020.

[35] ReNEWS. Vietnamese government inks La Gan survey contracts [EB/OL]. (2021 - 03 - 20) [2021 - 06 - 30]. https: //www. renews. biz/69722/vietnamese - government - signs - 35gw - la - gan - survey - contracts/.

[36] TALEB H M, HIJLEH B A. Optimizing the Power Generation of a Wind Farm in Low Wind Speed Regions [J]. Sustainability, 2021, 13 (9).

[37] VnExpress. Ministry calls for allowing private investment in power transmission [EB/OL]. (2020 - 05 - 04) [2021 - 06 - 30]. https: //e. vnexpress. net/news/business/economy/ministry - calls - for - allowing - private - investment - in - power - transmission - 4094268. html.